Italian Thought Today

This collection provides English readers with a critical update on current debates on biopolitics in and around Italian thought. More than a decade after the publication of seminal books such as Agamben's *Homo Sacer* and Hardt and Negri's *Empire*, the names of, among others, Roberto Esposito, Paolo Virno, Christian Marazzi, and Andrea Fumagalli have recently been brought to the attention of Anglophone scholars and political activists. Several authors have rightly emphasised the evanescent character of biopolitics, and the difficulty in providing a definition of it that could embrace all the conflicting theories of its most celebrated critics and supporters. The present collection is structured around the basic contention that bio-economy, human nature, and Christianity are the three visible contemporary manifestations of the theoretical object/problem of biopolitics in, respectively, Italian post-workerist economics, post-Marxist philosophical anthropology, and post-structuralist ontology.

This book was originally published as a special issue of *Angelaki: Journal of Theoretical Humanities*.

Lorenzo Chiesa is Professor of Modern European Thought and co-director of the Centre for Critical Thought at the University of Kent, UK. His most recent publications include *Subjectivity and Otherness* (2007), *The Italian Difference: Between Nihilism and Biopolitics* (2009) (co-edited with Alberto Toscano), and the English translation of Agamben's *The Kingdom and the Glory* (2011).

Italian Thought Today

Bio-economy, Human Nature, Christianity

Edited by
Lorenzo Chiesa

Routledge
Taylor & Francis Group
LONDON AND NEW YORK

First published 2014
by Routledge
2 Park Square, Milton Park, Abingdon, Oxon, OX14 4RN, UK

and by Routledge
711 Third Avenue, New York, NY 10017, USA

Routledge is an imprint of the Taylor & Francis Group, an informa business

British Library Cataloguing in Publication Data
A catalogue record for this book is available from the British Library

ISBN 13: 978-0-415-74874-2

Typeset in Garamond
by Taylor & Francis Books

Publisher's Note
The publisher accepts responsibility for any inconsistencies that may have arisen during the conversion of this book from journal articles to book chapters, namely the possible inclusion of journal terminology.

Disclaimer
Every effort has been made to contact copyright holders for their permission to reprint material in this book. The publishers would be grateful to hear from any copyright holder who is not here acknowledged and will undertake to rectify any errors or omissions in future editions of this book.

Contents

Citation Information

The chapters in this book were originally published in the journal *Angelaki: Journal of Theoretical Humanities*, volume 16, issue 3 (September 2011). When citing this material, please use the original page numbering for each article, as follows:

Chapter 1
Biopolitics in Early Twenty-First-Century Italian Theory
Lorenzo Chiesa
Angelaki: Journal of Theoretical Humanities, volume 16, issue 3 (September 2011) pp. 1–6

Chapter 2
Twenty Theses on Contemporary Capitalism (Cognitive Biocapitalism)
Andrea Fumagalli, translated by Sabrina Ovan
Angelaki: Journal of Theoretical Humanities, volume 16, issue 3 (September 2011) pp. 7–18

Chapter 3
Dyslexia and the Economy
Christian Marazzi, translated by Sabrina Ovan
Angelaki: Journal of Theoretical Humanities, volume 16, issue 3 (September 2011) pp. 19–32

Chapter 4
Hunger, Repletion, and Anxiety
Massimo Recalcati, translated by Wissia Fiorucci
Angelaki: Journal of Theoretical Humanities, volume 16, issue 3 (September 2011) pp. 33–38

Chapter 5
The Word and the Flesh: Postworkerism and the Biopolitics of Language in Paolo Virno and Christian Marazzi
Pietro Bianchi
Angelaki: Journal of Theoretical Humanities, volume 16, issue 3 (September 2011) pp. 39–52

Chapter 6
The Untamed Ontology
Davide Tarizzo, translated by Alvise Sforza Tarabochia
Angelaki: Journal of Theoretical Humanities, volume 16, issue 3
(September 2011) pp. 53–62

Chapter 7
The Anthropological Meaning of Infinite Regression
Paolo Virno, translated by Lorenzo Chiesa
Angelaki: Journal of Theoretical Humanities, volume 16, issue 3
(September 2011) pp. 63–76

Chapter 8
Politics and Human Nature
Roberto Esposito, translated by Lorenzo Chiesa
Angelaki: Journal of Theoretical Humanities, volume 16, issue 3
(September 2011) pp. 77–84

Chapter 9
Affirmative Biopolitics and Human Nature in Franco Basaglia's Thought
Alvise Sforza Tarabochia
Angelaki: Journal of Theoretical Humanities, volume 16, issue 3
(September 2011) pp. 85–100

Chapter 10
The Bio-Theo-Politics of Birth
Lorenzo Chiesa
Angelaki: Journal of Theoretical Humanities, volume 16, issue 3
(September 2011) pp. 101–116

Chapter 11
Angels
Giorgio Agamben, translated by Lorenzo Chiesa
Angelaki: Journal of Theoretical Humanities, volume 16, issue 3
(September 2011) pp. 117–124

Chapter 12
Divine Management: Critical Remarks on Giorgio Agamben's The Kingdom and the Glory
Alberto Toscano
Angelaki: Journal of Theoretical Humanities, volume 16, issue 3
(September 2011) pp. 125–136

Chapter 13
Giorgio Agamben's Godless Saints: Saving What Was Not
Jelica Šumič
Angelaki: Journal of Theoretical Humanities, volume 16, issue 3
(September 2011) pp. 137–148

Chapter 14
*Kafka's Land Surveyor K.: Agamben's Anti-*Muselmann
Boštjan Nedoh
Angelaki: Journal of Theoretical Humanities, volume 16, issue 3 (September 2011) pp. 149–162

Chapter 15
The Event of Language as Force of Life: Agamben's Linguistic Vitalism
Lorenzo Chiesa and Frank Ruda
Angelaki: Journal of Theoretical Humanities, volume 16, issue 3 (September 2011) pp. 163–180

Please direct any queries you may have about the citations to clsuk.permissions@cengage.com

Notes on Contributors

Giorgio Agamben is Baruch Spinoza Chair at the European Graduate School (EGS), former Professor of Theoretical Philosophy at the Istituto Universitario di Architettura of Venice, Italy, and former programme director at the Collège international de philosophie, Paris. He has also held visiting appointments at several American universities. His most recent books include *The Highest Poverty: Monastic Rules and Form-of-Life* (2013), *Opus Dei. An Archaeology of Duty* (2013), *The Kingdom and the Glory* (2011), *The Sacrament of Language* (2011), *Nudities* (2010), and *The Signature of All Things* (2009).

Pietro Bianchi was until recently Researcher in the Theory Department of the Jan Van Eyck Academy in Maastricht, The Netherlands and Visiting Researcher at the University of California – Los Angeles, USA. He holds a PhD from the University of Udine, Italy. His thesis focused on Jacques Lacan's reflection on the materialism of the object-gaze. He works and publishes on psychoanalysis, Marxism and visual studies.

Lorenzo Chiesa is Professor of Modern European Thought and Co-Director of the Centre for Critical Thought at the University of Kent, UK. His most recent publications include *Subjectivity and Otherness* (2007), *The Italian Difference: Between Nihilism and Biopolitics* (2009) (co-edited with Alberto Toscano), and the English translation of Agamben's *The Kingdom and the Glory* (2011).

Roberto Esposito is Professor of Theoretical Philosophy and Coordinator of the doctoral programme in Modern and Contemporary Philosophy at the Istituto Italiano di Scienze Umane (SUM) in Naples, Italy. He is also co-director of the journal Filosofia politica and the director of several book series for Italian publishers such as Laterza, Franco Angeli, and Bibliopolis. In the past, he served as the only Italian member of the international scientific committee of the Collège International de Philosophie, Paris, France. His most recent publications include *Bíos: Biopolitics and Philosophy* (2008), *Communitas: The Origin and Destiny of Community* (2009), *Third Person* (2012) and *Living Thought: The Origins and Actuality of Italian Philosophy* (2013).

Andrea Fumagalli is Associate Professor of Economics at the University of Pavia, Italy, and the Vice-President of the Basic Income Network–Italy. His research interests include cognitive capitalism, bioeconomy, labour transformations and the theory of money. The English translation of his *Crisis in the Global Economy* (edited with Sandro Mezzadra) was released in 2011. Among his other recent publications are two seminal books for contemporary debates on postworkerism: *Bioeconomia e capitalism cognitivo* (2007) and *La moneta nell'Impero* (2002) (edited with Christian Marazzi and Adelino Zanini).

Christian Marazzi is currently Director of Socio-Economic Research at the Scuola Universitaria della Svizzera Italiana, Switzerland. He previously taught at the University of Padova in Italy, the State University of New York, USA, and at the University of Lausanne. Semiotext(e)/MIT Press published the English translation of four of his books: *The Violence of Financial Capitalism* (2010), *Capital and Language* (2008), *Autonomia* (edited with Sylvère Lotringer) (2007) and his seminal work on the post-Fordist economy, *Capital and Affects* – first published in 1994 (2011).

Boštjan Nedoh is a Research Fellow at the Institute of Philosophy, Scientific Research Centre of the Slovenian Academy of Science and Art in Ljubljana, Slovenia. His main research interests are in the area of contemporary continental philosophy, Lacanian psychoanalysis, and Marx's critique of political economy. He has published articles on these topics and translated essays by Agamben into Slovenian.

Massimo Recalcati is a practising Lacanian Psychoanalyst and the Founder of JONAS, Centre for the Psychoanalytic Clinic of New Symptoms. He teaches Psychopathology of Eating Behaviours at the University of Pavia, and Sociology of Collective Phenomena at the Università Cattolica del Sacro Cuore, Milan. His many books include *L'ultima cena: anoressia e bulimia* (1997), *Jacques Lacan: un insegnamento sul sapere dell'inconscio* (2000), *Clinica del vuoto: anoressie, dipendenze, psicosi* (2002), *Lo psicoanalista e la città: l'inconscio e il discorso del capitalista* (2007), and *L'uomo senza inconscio. Figure della nuova clinica psicoanalitica* (2010).

Frank Ruda is a Research Associate in the Department of Philosophy at the Free University of Berlin, Germany, co-editor of the *Morale Provisoire* book series at Merve, and one of the initiators of the Materialism and Dialectics collective. He is the author of *Hegels Pöbel* (2011) and has published extensively on Continental philosophy and Marxism. He has also translated books by Alain Badiou and Jacques Rancière into German.

Alvise Sforza Tarabochia is a Lecturer in Italian Studies at the University of Kent at Canterbury, UK. He is the author of the monograph *Psychiatry, Subjectivity, Community. Franco Basaglia and Biopolitics* (Peter Lang, 2013), member of the Centre for Critical Thought at the University of Kent, and also a member of the editorial board of the *European Journal of Psychoanalysis*. He has recently edited an issue of the journal *EJPsy*, entitled *The End* (2014).

Jelica Šumič is Senior Research Fellow at the Institute of Philosophy, Centre for Scientific Research Ljubljana, Slovenia, and Professor of Contemporary Philosophy at the University of Nova Gorica, Slovenia. She has been a Visiting Professor at the University of Paris 8, France, University of Essex, UK, University of San Martin, Buenos Aires, and Northwestern University, USA. She conducted a seminar on Lacan's politics at the Collége international de philosophie, Paris. She is the author of *Mutations of Ethics* (2002), *De l'Autre au Même* (2011), *Ethics of Silence* (2011), and *Volonté et désir* (2011). She has edited and contributed to the volume *Universel, Singulier, Sujet* (2000) (with Alain Badiou et al.), and the international issues of the journal *Filozofski Vestnik*, namely *Philosophie, psychanalyse: alliance ou misalliance*

(2006); *What is to Live?/Qu'est-ce que vivre?* (2009); and *Les Pouvoirs des desires indicibles* (2010) (together with Jean-Pierre Marcos).

Davide Tarizzo teaches Moral Philosophy at the University of Salerno, Italy, and Political Philosophy at the University of Naples "L'Orientale". He is also a scientific consultant to the PhD programme in Philosophy of the Istituto Italiano di Scienze Umane. He has edited the Italian translation of several books by contemporary philosophers such as Gilles Deleuze, Jean-Luc Nancy, Alain Badiou, Ernesto Laclau, and others. Among his most recent publications are: *Il pensiero libero. La filosofia francese dopo lo strutturalismo* (2003), *Introduzione a Lacan* (2003), *Giochi di potere. Sulla paranoia politica* (2007), and *La vita, un'invenzione recente* (2010). He is currently writing a book on the modern political notion of the "people".

Alberto Toscano teaches in the Department of Sociology at Goldsmiths, University of London, UK. He is the author of *Fanaticism: On the Uses of an Idea*, and an editor of the journal *Historical Materialism*.

Paolo Virno is Professor of Philosophy of Language at the University of Rome, Italy. He is the author of several books including *A Grammar of the Multitude* (2002) and *Multitude. Between Innovation and Negation* (2008). His most recent publications in Italian are *Saggio sulla negazione. Per un'antropologia linguistica* (2013), and *E cosí via, all'infinito. Logica e antropologia* (2010).

lorenzo chiesa

BIOPOLITICS IN EARLY TWENTY-FIRST-CENTURY ITALIAN THEORY

The main aim of this collection of essays is to provide English readers with a critical update on current debates on biopolitics in and around Italian thought. More than a decade after the publication of seminal books such as Agamben's *Homo Sacer* and Hardt and Negri's *Empire* the contemporary scene is incredibly more complex than is usually assumed. On the one hand, the works of these prolific authors continue to be translated at an incredible rate and still stand firmly at the centre of most discussions concerning the redefinition of radical international thought at the time of an alleged global "War on terror," the concomitant identification of a new figure of enmity, but also the resurgence of a "communist hypothesis" that resists such an ideology. On the other hand, several new names have recently been brought to the attention of anglophone scholars and political activists: in the last few years, major American and British publishing houses have released volumes by thinkers such as Esposito, Virno, Marazzi, and Fumagalli.

One of the basic presuppositions of this collection is that these more recent works have so far been received through an interpretative lens that is to some extent obsolete. For instance, Esposito's work is all too often relegated to occupying a sort of median position between the supposedly opposed readings of biopolitics – negative and affirmative, respectively – advanced by Agamben and Negri, and to a doxastic, parochial and ultimately sterile problematisation of the Foucauldian understanding of the relation between biopower and sovereignty. What this misses out is Esposito's courageous return to the much broader theoretical question concerning human nature and the human animal. Despite post-structuralism's and deconstruction's extensive critique of the notion of "man" and its humanist biases, early twenty-first-century Italian thought is experiencing a resurgence of interest in this classical philosophical issue. Esposito's project (but also, in different guises, Virno's and Agamben's) proposes not only an enquiry into the relation of dependence between different figures of the subject and their material substratum, or into the ways in which subjectivity opens up an unsurpassable gap in nature, but also and especially an identification of the subject as irreducible to nature with its own animality. As witnessed by Tarizzo's essay, this issue is itself inextricable from a more general survey of the notion of life and the metaphysical and theological – Christian – biases that still permeate both its speculative and scientific understanding.

Similarly, the philosophical and economical work of the so-called post-workerists has so far, for the most part, been confined to the vague, almost eschatological, Negrian notions of multitude and exodus,[1] thus omitting to emphasise their timely analysis of the new economy and the financial crises dictated by cognitive capitalism (as is the case with Marazzi and Fumagalli) or their persistent interaction with analytic philosophy, formal logic and cognitive sciences (as is the case with Virno and other thinkers collaborating with him).[2] Having said this, the present collection also starts off with the assumption that these reductive readings are to a certain extent promoted by the very authors in question. For example, as shown by Bianchi's essay, it is indeed the case that both economical and philosophical post-workerism rely on the quasi-apocalyptic assumption that the current phase of capitalism is truly exceptional and irreversible – to put it bluntly, a certain extreme notion of epochality, if not of the end of history, is taken for granted in so far as the contemporary form of accumulation is deemed to cancel the distinction between biology and history. It is also incontrovertible that Esposito's own insistence on a "third way" for biopolitics, epitomised by the inextricability of *immunitas* and *communitas*, runs the risk of falling back into a pre-Foucauldian metaphysics of life and man that his very insights into a truly materialist overcoming of the divide between human and natural sciences attempt to refute.

This special issue does not claim to be exhaustive. Essays by other significant authors, who are as yet mostly unknown outside of Italy, have not been included for reasons of space.[3] Most importantly, we must stress that discussions on biopolitics are far from being hegemonic in contemporary Italian academia; with specific regard to philosophy, in the last fifteen years we have rather increasingly witnessed the strengthening of a watered-down adaptation of Anglo-American analytic speculation and the more than dubious transformation of vast sectors of "weak thought" (in short, the local version of postmodernism that prospered throughout the 1980s and 1990s) into *consulenza filosofica* (that is, the bleak idea that the contemporary task of the philosopher is to act as a consultant to businessmen and corporations).[4] The decision to call a patent minority by the name of "Italian thought" can nevertheless be motivated by the fact that it is precisely the investigation into biopolitics that has established the names of a number of Italian thinkers worldwide, especially in the last decade, and that, paradoxically, what remains at an international level a predominantly academic, and hence restricted, phenomenon has started being broadly covered in Italy by journalists and cultural analysts from major national newspapers which are usually indifferent to theoretical debates and innovations.[5] Without necessarily going as far as arguing that the link between life and politics has always been the privileged target of Italian philosophy – from Machiavelli to Croce, Bruno to Gentile, Vico to Pasolini[6] – we can cautiously suggest that, perhaps more than any other speculative European tradition, Italian thought has time and again been able to connect theory with praxis, as well as be truly open to other disciplines, in ways that have given rise to unforeseeable short-circuits. The latter are very diverse and include, for instance, not only the hermeneutic possibility of convincingly revisiting Gramsci's thought by means of the concept of bio-economy in order to prove its topicality[7] but also, more concretely, in the domain of current affairs, that of a prime minister adopting and distorting Pasolini's idea of "anthropological difference" to distinguish between right-wing and left-wing Italians and attack the judges who intend to prosecute him, or conversely, that of a prominent Lacanian psychoanalyst writing and speaking on television about the same prime minister as the paradigm of a new post-human epoch of "monadic enjoyment."[8]

The contributors to this collection of essays are active in at least five different official disciplines: philosophy, economics, sociology, psychoanalysis, and Italian Studies. When dealing with biopolitics, I would argue that we should understand interdisciplinarity on the basis of Althusser's theory of discourses, that is, not as an "'interdisciplinary theme', but [as] a theoretical object, a fundamental theoretical problem which, while it may well touch on the domains of several existing disciplines, will not necessarily appear in person in any of them."[9] Several authors have rightly emphasised the evanescent character of biopolitics, the difficulty in providing a definition of it that could embrace all the conflicting theories of its most celebrated critics and supporters. The present collection is structured around the basic contention that bio-economy, human nature, and Christianity are the three visible contemporary manifestations of the theoretical object/problem of biopolitics in, respectively, Italian post-workerist economics, post-Marxist philosophical anthropology, and post-structuralist ontology. It is around these areas of research that original investigations belonging to the disciplines mentioned above seem to coalesce and almost fuse.

The first two essays of this issue share the bio-economic thesis that contemporary capitalism directly extracts value from the generic faculty of language, the invariant species-specific features of the human animal. While Fumagalli's programmatic manifesto outlines the way in which the contemporary form of capitalism may be defined as a "cognitive biocapitalism," Marazzi's text focuses in detail on dyslexia, which he considers to be a symptomatic expression of the digital dimension of the same economy. On his part, in the third essay, Recalcati moves from a contiguous socio-economic assessment to psychoanalytically examine anorexia as an emblematic consequence of the structural illusions promoted by the contemporary discourse of the capitalist. Bianchi's essay then closes the first section dedicated to bio-economy by challenging from a more classical Marxist perspective the post-workerist substantialist assumption – present in both Marazzi's investigation of bio-economy and Virno's political philosophy – that there is such a thing as a "common behaviour of mankind" which current capitalism would immanently control and put to work.

The second group of essays opens with Tarizzo's analysis of how our scientific notion of "life" is still biased by metaphysical assumptions and, more specifically, by a vitalist ontology that regards life as a "secret force" and thus strongly influences our comprehension of the human animal, especially after the biopolitical turn. The two contributions that follow, by Virno and Esposito respectively, converge on the claim that a proper understanding of *Homo sapiens* as a political animal should overcome the spurious divide between natural and social sciences. On the one hand, Virno scrutinises the anthropological meaning of the logical category of infinite regression – to be seen as a paradigm of what he calls the "naturally artificial disposition of the human animal" – within the more general framework of a new materialist Marxist anthropology that remains faithful to the theory of evolution. On the other hand, Esposito returns to the question of humanism, in particular twentieth-century philosophy's failed attempt to dispose of it, and concludes that the overbearing entrance of biological life into socio-political dynamics is not necessarily a danger from which we have to protect ourselves in the name of a self-centred purity of the individual and the species. In the concluding essay of this section, Sforza Tarabochia unfolds precisely this aspect of Esposito's thought by taking into account Basaglia's anti-psychiatric notion of human nature and the possibility of developing an affirmative biopolitical psychiatry that would continue his successful dismantling of the old mental health disciplinary apparatus.

The third and final section of this collection is arguably the most critical. Chiesa's essay considers Esposito's notion of birth as an affirmative biopolitical category and shows that, in spite of its originality and the possibility it offers to dismantle the thanatopolitical drift of Pasolini's considerations on abortion, it still implicitly presupposes the taking of a stance on the transcendent differentiality of life and the legal obligation towards it that closely follows the dogmas of the Roman Catholic Church. In similar fashion, Toscano uncovers the shortcomings of Agamben's recent work on the theological genealogy of the economy and government,[10] especially regarding his notion of failed secularisation, which Toscano thinks is the consequence of a Christian historical substantialism that clashes with any claim to be engaging in a genealogy. Agamben's own contribution responds to Toscano's critique by expanding on the genealogy of the link between the modern notion of government and angelology that he carried out in *The Kingdom and the Glory* by means of a comparative analysis of this connection in Judaism, Christianity, and Islam. His conclusion that "the government of the world is still today in the hands of the Christian Occident" since "Christianity is the only one of the three monotheistic religions that has turned the government of the world into an internal articulation of divinity and has, thus, divinised angelic power" will no doubt ignite new debates. For her part, Šumič dwells on Agamben's treatment of Christian theology, paying particular attention to the themes of salvation and of the messianic subject; after deconstructing his position, she agrees with him in identifying the truly emancipative and redemptive kernel of politics in the act that "decreates the decreation," which she associates with both the figure of Bartleby as – in Deleuze's own words – "the new Christ," and the practice of Lacanian psychoanalysis.

The two remaining essays should be regarded as divergent attempts to contextualise Agamben's essay and the articles on his appropriation of Christianity contained in the third section within his overall ontology and notion of political agency.[11] While Nedoh maintains that, in spite of the evident limitations of the Agambenian theory of the – essentially passive – subject, we can detect a sort of Agamben beyond Agamben in the character of land surveyor K. from Kafka's *The Castle* (of which the Italian philosopher has recently offered a brilliant interpretation), Chiesa and Ruda dispute this and identify the ultimate embodiment of a bio-theo-politics sutured to a vitalist ontology of language in the unsettling figure of the pervert.

It goes without saying that the tripartite division I adopt to introduce this collection is far from being rigid. Most essays explicitly deal with more than one of the general themes that I refer to in the title of the collection. Agamben's and Chiesa's contributions from the third section talk as much about Christianity as about bio-economy and human nature, while the articles by Marazzi and Fumagalli from the first do not deal with the question "what is the speaking being?" any less extensively than does Esposito's piece from the second. More importantly, readers are invited to make these essays dialogue in alternative ways and, in so doing, increase their awareness of the multiple "interdisciplinary" bridges that unify the field of the theoretical object/problem that contemporary Italian thought calls "biopolitics." Thus, for example, Sforza Tarabochia's considerations regarding the dangers or emancipative potentiality of a biopolitical approach to psychiatry echo Recalcati's psychopathology of everyday life at the time of the full disclosure of a form of capitalism increasingly based on a consumerism that directly targets the very life of the species. This in turn reiterates the centrality of the identification of consumption and production found in key passages of Fumagalli's, Marazzi's, and Chiesa's contributions,[12] which then resonates with Tarizzo's warning against reading these social phenomena through the filter of a metaphysics of life, and so on.

notes

1 This is confirmed by the deceptive English title – *Multitude: Between Innovation and Negation* (Cambridge, MA: Semiotext(e), 2008) – given to a recently published collection of Virno's essays which deals only in part and diagonally with the question of the multitude. The main topics of the book are, quite tellingly, the notion of evil, jokes, and mirror neurons.

2 A clear example of this is the journal *Forme di vita*, edited by Virno, De Carolis, Cimatti, and Catucci (as well as, initially, Agamben) between 2003 and 2007, which regularly featured interviews with leading cognitive scientists and articles on and by, among others, Wittgenstein, Chomsky, and Bolk. The editorial of its sixth issue was significantly entitled "In Defence of Analytic Philosophy."

3 Other noteworthy Italian authors who work on biopolitics and related issues include Laura Bazzicalupo in political philosophy (*Il governo delle vite. Biopolitica ed economia* (Rome and Bari: Laterza, 2006)), Massimilano De Carolis in

philosophical anthropology (*Il paradosso antropologico* (Macerata: Quodlibet, 2008)), Franco Lo Piparo in philosophy of language (*Aristotele e il linguaggio* (Rome and Bari: Laterza, 2003)), and Roberto Marchesini in ethology and zooanthropology (*Post-Human. Verso nuovi modelli di esistenza* (Turin: Bollati Boringhieri, 2002)).

4 For a sociological demolition of this widespread movement, see A. Dal Lago, *Il business del pensiero. La consulenza filosofica tra cura di sé e terapia degli altri* (Rome: Manifestolibri, 2007). The – to say the least alarming – postmodern justification usually advocated by the promoters of this practice is hinted at by the very title of the book, "cura di sé" being the Italian translation of the late Foucault's notion of "souci de soi" ("care of the self" in English).

5 See, for instance, M. Pasquinelli, "L'ascesa in cattedra di un pensiero critico," *Il manifesto* 13 Apr. 2011. Alberto Toscano and I made a similar choice concerning the title of our *The Italian Difference: Between Nihilism and Biopolitics* (Melbourne: Re.press, 2009).

6 Esposito cogently develops this argument in his latest book, *Pensiero vivente. Origine e attualità della filosofia italiana* (Turin: Einaudi, 2010).

7 See ibid. 178–91.

8 See "Ci sono magistrati eversivi trascinati da sinistra criminale comunista," *La Repubblica.it* 16 Apr. 2011, and M. Recalcati, *L'uomo senza inconscio. Figure della nuova clinica psicoanalitica* (Milan: Cortina, 2010). See also M. Recalcati, "Mutazioni antropologiche e berlusconismo," available <http://www.youtube.com/watch?v=UdXfOn7r3fY>, and the interview he gave in early 2011 to the well-known political talk-show *L'infedele*.

9 L. Althusser, "Three Notes on the Theory of Discourses" in *The Humanist Controversy and Other Writings* (London: Verso, 2003) 33.

10 Esposito has himself dedicated a short section of *Pensiero vivente* to a persuasive critique of *The Kingdom and the Glory* and of what he calls Agamben's "belonging to the same theological paradigm that he intends to defuse" (*Pensiero vivente* 243).

11 The fact that four essays of the present collection are devoted to an assessment of Agamben's work proves that his writings continue to stimulate discussions.

12 The identification between production and consumption in biopolitical capitalism has also been investigated from a different – Arendtian – perspective by Bazzicalupo in *Il governo delle vite*, especially 118–25.

andrea fumagalli

translated by sabrina ovan

TWENTY THESES ON CONTEMPORARY CAPITALISM (COGNITIVE BIOCAPITALISM)

foreword

The aim of this article is twofold. On the one hand, it is an attempt at systematizing a series of reflections and concepts elaborated by a number of studies that appeared in the last decade. This research comes from scholars in different disciplines, but who identify, even in their internal differences, with a method of analysis rooted in the Italian Workerist thought of the 1960s. For this reason, it is a work in progress, and has no pretense of being exhaustive.[1] On the other hand, it ambitiously tries to communicate and clarify an issue that has provoked much debate in the last few years, especially in the field of heretic and heterodox thought, that is to say, the analysis of the salient characteristics of the current state of capitalism.

From the very title, we formulate a thesis: the contemporary form of capitalism is defined in a univocal way as *cognitive biocapitalism*. The twenty theses that follow are a means of justifying this definition.

In the last thirty years, the current process of capitalist accumulation and valorization has assumed different names: the most common of these, *post-Fordism*, is also the oldest. The term post-Fordism became popular during the 1990s, especially through the French *école de la regulation*.[2] The term, however, is not without its ambiguities and diverse interpretations, as are all terms defined in a negative way. Our idea is that with the term *post-Fordism* we define the period, from the 1975 crisis to the early 1990s crisis, during which the process of accumulation and valorization is no longer based on the centrality of Fordist material production, the vertically integrated, large factory. At the same time, in this period, we do not yet possess an alternative paradigm. Unsurprisingly, in the prefix "post-," we express what is no longer there, without underlining what actually appears in the present. The post-Fordist phase is in fact characterized by the conjoined presence of more productive models: from the Japanese Toyotist model of the "just in time" derived from Taylorism[3] to the industrial district model of small enterprises[4] and the development of productive lines that tend to become international according to a hierarchy.[5] Among these models, it is still impossible to identify a hegemonic paradigm.

After the first Gulf War, innovations in the fields of transportation, language and communication (ICT) started to gather around a new single paradigm of accumulation and valorization. The new capitalist configuration tends to identify in "knowledge" and "space" (geographic and virtual) as commodities a new foundation for dynamic skills of accumulation. As a consequence, two new dynamic economies of scale are formed, which are the basis for the growth in productivity (or, the source of surplus value): learning economies and network economies. The first are connected to the process of generation and creation of new knowledge (based on new systems of communication and information technologies); the second derive from the organizational modalities of each district (territorial networks or system areas), which are no longer used for production and distribution only, but increasingly as a vehicle of diffusion (and control) of knowledge and technological progress. We can name this paradigm of accumulation *cognitive capitalism*:[6]

> The term *capitalism* designates the permanence, though metamorphic, of the fundamental variables of the capitalistic system: the leading role of profit, and the wage system in particular, or more precisely, the different forms of employed labor from which surplus value is extracted. The attribute *cognitive* evidences the new nature of labor, of the sources of valorization and property structure, on which the process of accumulation is founded, and the contradictions that this mutation generates.[7]

The centrality of learning and network economies, typical of cognitive capitalism, is put into question at the beginning of the new millennium, following the bursting of the Internet economy bubble and its speculations, in March 2000. The new cognitive paradigm alone is unable to protect the socio-economic system from the structural instability that characterizes it. It is also necessary for new liquidity to be directed into the financial markets. The ability of financial markets to generate "value" is tied to the development of "conventions" (speculative bubbles) which can create somewhat homogeneous expectations, thereby pushing the main financial operators to support certain types of financial activities.[8] What the Internet economy did in the 1990s was followed in the 2000s by the great attraction to the development of Asian markets (China entered the WTO in December 2001) and real estate. Today, the focus is mostly on the performance of European welfare. Independently of the dominant convention, contemporary capitalism is always in search of new social and vital circles to absorb and commodify, involving more and more the bare vital faculties of human beings. It is for this reason that in the last few years we have been hearing about *bioeconomy* and *biocapitalism*.[9]

At this point, the reader should clearly understand how the term used in these pages is nothing but the contraction between cognitive capitalism and biocapitalism: *cognitive biocapitalism* is the phrase that defines contemporary capitalism.

Thesis #1: in cognitive biocapitalism, the financial markets, knowledge and relations are the motor of accumulation.

Financial markets are the pulsating heart; knowledge is the brain; relational activities are the nervous system. Cognitive biocapitalism is a single body, inside of which the "real" sphere cannot be separated from the "financial," nor can the productive sphere be separated from the unproductive, or work-time from life-time, or production from reproduction and consumption...

Thesis #2: in cognitive biocapitalism, financial markets directly influence and condition the process of accumulation and valorization.[10]

In a broader sense, financialization marks the definitive passage from *commodity money* to *sign money*.[11] With the complete dematerialization of money (after the Bretton Woods crash in 1971, marking the end of the convertibility of the dollar to gold), financial markets define the social and hierarchic conventions which are able to secure short-term monetary value. At the same time, they leave open the relations of debit and credit, provided sufficient trust is generated in the operators. From this viewpoint, financial markets lubricate the process of accumulation. In the capitalistic system, in fact, there is no accumulation without debt. It is no coincidence that, from the 1990s onward, financial markets have taken care of financing accumulation activities: the liquidity drawn by financial markets rewards the restructuring of production aimed at exploiting knowledge and the control of spaces external to the enterprise. Secondly, in the presence of surplus value, financial markets have the same role in the current economic system that the Keynesian multiplier (activated by deficit spending) had in industrial-Fordist capitalism. However – unlike the classic Keynesian multiplier – the new financial multiplier leads to a distorted redistribution of revenues. For such a multiplier to be operative (>1), the financial basis (that is, the extension of financial markets) must be constantly growing, and the capital gain must be, on average, higher than the median salary loss. On the other hand, the polarization of revenues increases the risk of debt insolvency, which is the basis of the growth of the very financial foundation, and reduces the median salary. Thirdly, financial markets, forcibly channeling growing portions of work revenues (such as severance indemnity and social security, as well as earnings that, through the social state, turn into institutions for health and public education), substitute in this way the state as a social provider. From this point of view, financial markets represent the privatization of the reproductive sphere of life. Finally, financial markets are the place where today capitalistic valorization is established, that is, the place where the exploitation of social cooperation and of the general intellect is measured by way of the dynamic of stock market values. As a consequence, profit transforms into rent (see Thesis #3), and financial markets become the place where labor-value is determined and transformed into *finance-value*. The latter is nothing other than the subjective expression of the expectation of future profits articulated by financial markets, which in this way secure a rent.

Financial markets thus exercise biopower.[12]

Thesis #3: in cognitive biocapitalism, we register the becoming-rent of profit.[13]

Rent is the main capturing tool of both surplus value and the de-socialization/privatization of what is common. The meaning and the key role of this becoming-rent of profit can be appreciated at two levels.

On the one hand, this process is evident at the level of the social organization of production and of the distribution of revenues: the criteria underlying the traditional distinction between profit and rent become less and less pertinent. The confusion of the frontiers between rent and profit finds one of its expressions in the way in which financial power remodels the very criteria of company governance under the sole aim of creating value for the shareholder. In cognitive biocapitalism, not only do we witness the final decline of the Weberian entrepreneur (the figure combining the functions of ownership and direction of the firm, who had already partly disappeared in industrial-Fordist capitalism after the marginalist revolution of the 1930s). We also see the irreversible crisis of the Galbraithian techno-structure, legitimized in its role by the planning of innovation and the organization of labor. The new governance of today's companies is increasingly founded on a type of management whose principal competence is exercising financial and speculative functions, while delegating to employed labor the real functions of the organization of production.

On the other hand, the competitiveness of a company is largely dependent not on internal economies but on external ones, that is to say, on the ability to capture productive surpluses that come from the cognitive resources of a territory. Capital, then, does benefit freely from the collective knowledge of society, as if it were a "gift of nature." From this point of view, the becoming-rent of profit takes the form of a privatization of what is common,[14] gaining revenues from the creation of a scarcity of resources that is only artificial. It is the common that links together, in a single logic, the rent coming from real estate speculation and financial rent – which, since the beginning of the 1980s, played a major role in fiscal crisis and the dismantling of welfare state institutions, as a result of the privatization of currency and public debt. The becoming-rent of profit derives, then, from the attempt at privatizing knowledge and life (*bios*). This is achieved thanks to a politics promoting the reinforcement of intellectual property rights so that the cost of numerous commodities is kept artificially high, although their reproduction costs are extremely low or even close to zero.

Thesis #4: in cognitive biocapitalism, value production is no longer founded on material production alone.

Productive activity is increasingly based on immaterial elements, that is to say, on intangible "raw materials," which are very hard to measure and quantify, and which come directly from the utilization of the relational, sentimental, and cerebral faculties of human beings. The process of valorization loses, in this way, the measuring unit usually connected to material production. This measure used to be somewhat defined according to the necessary amount of labor needed for the production of commodities, measurable on the basis of the tangibility of production and the necessary time for production. With the advent of cognitive capitalism, valorization tends to graft itself onto different forms of labor, which go beyond the official work-time and coincide more and more with the whole life-time. Today, the value of labor at the basis of biocapitalistic accumulation is also the value of knowledge, of affects and relationships; it is the value of the imaginary and the symbolic (cf. Thesis #15).

Thesis #5: in cognitive biocapitalism, value production is no longer founded on a homogeneous, standardized scheme for the organization of labor, independently of the type of goods produced.

The activity of production is carried out with different organizational modes, which are characterized by a network structure, thanks to the development of technologies for linguistic communication and transportation. What follows is a disruption of the traditional and unilateral hierarchic form typical of the factory. This is substituted by hierarchic structures activated on the territory along sub-supply production chains, and characterized by relations of cooperation and/or control.

Thesis #6: in cognitive biocapitalism, the division of labor takes on itself cognitive characteristics, and therefore is based on the differential access and use of different forms of knowledge.

Knowledge can be divided into four levels: *information*, *codified knowledge*, *tacit knowledge* and *culture* (or *systemic knowledge*), characterized by unilateral relations of dependence. *Information* is the basic level of knowledge that is more and more incorporated into the machine element. *Codified knowledge* is a specialized knowledge (a know how) that derives from tacit knowledge but that is transmitted through standardized procedures, with machines as intermediary, as a consequence of which its bearer can be substituted at any moment, having no contractual power. *Tacit knowledge* can derive from personal learning processes or from specific investments in R&D (thanks to intellectual property rights); furthermore, at least until codified, it can only be transmitted through a human being, thus possibly generating forms of *enclosures*. Those who possess tacit knowledge, which is relevant for the productive process, therefore have a high contractual power and define the hierarchical structure of labor and production. However, tacit knowledge, if relevant, is destined to transform into codified knowledge, sooner or later, and thus lose value. Lastly, *culture* is the set of knowledges that allows one to hold the *intellectual function*, that is to say, the ability to act critically and creatively, not immediately subsumed to the logic of biocapitalist valorization. As a consequence, culture is dangerous for the reproducibility of the socio-economic system, and it constitutes also a surplus that exceeds control.

Thesis #7: in cognitive biocapitalism, the condition of the labor force goes hand in hand with mobility and the predominance of individual contracting (*precariousness*).

This derives from the fact that nomadic individualities are put to work, and the primacy of private rights over workers' rights brings about a transformation of the contribution of individualities, especially if characterized by cognitive, relational and affective activities, into contractual individualism. Work relations based on precarious conditions, that is to say, the temporal limit and spatial mobility of labor, are the basic paradigm in which the relationship between capital and labor takes place. Precariousness then becomes a structural, existential and generalized condition.

Thesis #8: in cognitive biocapitalism, the accumulation process is founded on the exploitation of two new types of scale economies – the dynamic processes of learning and the dynamic processes of networking.

If knowledge is the basis of accumulation, it becomes unavoidable to analyze how its exchange and diffusion affect the dynamics of productivity. The peculiarity of cognitive biocapitalism is its ability to enlarge both knowledge-learning processes and network economies. Learning economies depend on the degree of *cumulativeness*, *opportunity* and *appropriability*.[15] Here, *opportunity* is defined as the expected rate of profit and, therefore, the higher the expected profit in adopting a new technology, the higher is the speed of its diffusion. *Cumulativeness* and *appropriability* represent the capacity of new knowledge to generate further innovation whilst avoiding the possibility of its imitation, thanks to the existence of intellectual property rights. Network economies depend on the level of income and positive externalities (E). When *learning economies* are constrained by intellectual property rights, we shall see that the consequence is that the greater the degree of *appropriability* of knowledge, the smaller becomes its capacity of diffusion – affecting, de facto, its ability to positively influence the associated productivity.[16] Whilst it is during the learning process that the generation of knowledge occurs, network economies define the way in which the produced knowledge is diffused. In a social system geared around innovation and production, investment policies depend upon R&D and "learning by doing" strategies and processes. In cognitive biocapitalism the impact of new Information and Communication Technologies based on computer science, micro-electronics and the new organizational productive changes (e.g. just-in-time, zero stock) have sped up the "learning by doing" processes, spreading them well beyond the firm. At the same time, part of the R&D process unfolds within territories each having one or more specific competencies. Where to locate economic activities is determined mainly by the search on the part of the firm for advantages in the development of its competencies.[17] Consequently, the productivity entailed by the exchange of knowledge cannot be assimilated to material productivity.

Thesis #9: an essential character of cognitive biocapitalism is the dematerialization of fixed capital, and the transfer of its productive and organizational functions to the living body of labor-power.

This process lies at the origin of one of the paradoxes of new capitalism: the contradiction between the rise in importance of cognitive work as a lever for the production of wealth and, at the same time, the devaluation of that work as far as salary and the profession are concerned. This paradox is inherent in Marazzi's definition of "the anthropogenetic character of contemporary capitalistic production," underlined in one of his essays.[18] In cognitive biocapitalism, the living being contains within itself the functions of both fixed and variable capital, that is, of both the material and machinery forms of labor belonging to the past and of the living labor of the present: bios.

Thesis #10: in cognitive biocapitalism, the separation between abstract labor and concrete labor is not as clear as it was in industrial-Fordist capitalism.

First of all, what Marx used to call "concrete labor," or labor producing use value, can be renamed today *creative labor*.[19] This term allows us to better understand the cerebral contribution inherent in such activity, while the term "concrete labor," though being conceptually its synonym, refers more to the realm of "making" than to that of "thinking," with a closer allusion to craftsmanship proper.

Thesis #11: in cognitive biocapitalism, we see more and more an interpenetration between place of production and productive networks.

Space, be it geographic or virtual, becomes a place of production no longer characterized by a unique and self-centered presence but rather by an ensemble of polycentric formal and informal networks. Production is the result of a flux structure, and such flux is always more immaterial or redesigned and directed by immaterial networks, especially when the commodities produced are material. A flux structure presupposes the centrality of linguistic networks of communication and the development of social cooperation. Such cooperation involves both the transmission of symbols and the logistical transportation of commodities and goods. Within this space, however, cooperation, which is far from being horizontal, develops along new trajectories of spatial partition of production, and cognitive division of labor. Reticular production – the network – is, in other words, a molecular space, and as such it is individualized, characterized by individual relations that most of the time produce cooperation in the end, but are not necessarily cooperative with one another.

Thesis #12: in cognitive biocapitalism, commodities have new meanings.

The value of commodities is no longer definable only along the lines of "necessary work-time." To that value, which does not disappear, another value is added, which derives from the degree of social symbolicity [*simbolicità*] that the commodity contains. The symbolic value of commodities increases in direct relation to their level of immateriality. It is in this field that the relation between production and realization (consumption) of commodities is played out. Not only does consumption realize the value of commodities, but it valorizes them at the same time.[20]

Thesis #13: in cognitive biocapitalism, life itself becomes value.

The theory of the value of work becomes a theory of the value of life.[21] This happens through the valorization of the differences that individuals possess. These differences, in their uniqueness, make possible the relational activities that are the basis of the social cooperation producing *general intellect*. Beside the general differences based on race, gender, and so on, we need to add up differences *tout court*, which are valorized without any relation to the anthropological characteristics that define them. What is now starting to be segmented and divided are cerebral differences, that is to say, individualities. Spatial and biologic differences, gender and race in particular, can at most be instruments for the immediate disciplining of the social body. The preoccupying emerging tendency, however, is the constitution of a human subjectivity characterized by the contradictory conflict between creative actions and cerebral standardization: the creation of a sort of bionic being, capable of managing the anthropogenetic process of production. These elements suggest a world where individuality is erased, but individualism is exalted. Cognitive biocapitalism is bioeconomic production: it is *bioeconomy*.

Thesis #14: in cognitive biocapitalism, differences become value[22] (see Thesis #13)

The traditional binary dichotomies inherited from industrial-Fordist capitalism are no longer topical. We are witnessing the overcoming of the *separation between life-time and work-time*. As soon as work activities use the vital faculties of individuals, it becomes impossible to define a temporal barrier between work- and non-work-time. Even if this distinction can nominally continue to exist on a formal-juridical level, the difference between life and work no longer exists in reality, and this is also due to the new language and communication technologies. Life appears completely *subsumed* into work. We are also witnessing the overcoming of the *separation between workplace and life-space*. The multiple forms of bio-labor are in fact *nomadic labor*, where mobility is a primary requisite. This phenomenon leads to the definition of *non-places of work*, as opposed to classic forms of *domestication*. In this case, indeed, we should not talk about a coincidence between workplace and life-space, but rather about the expropriation of the workplace, and of all possible consequences that this process might have on work identity. We are witnessing the overcoming of the *separation between production and reproduction*. This is the first consequence of life becoming work. When we talk about life, we do not only mean it as directly finalized to productive activity, but also to the social reproduction of life itself – a clear example of which is the almost exclusively female caretaking work. Having said this, we can state that the erasure of this distinction implies the partial overcoming of the specific gender difference, and poses the question of *differences tout court*.[23] In conclusion, we are witnessing the overcoming of the *separation between production, circulation, and consumption.* In cognitive biocapitalism, the act of consumption is, at the same time, a participation of public opinion, an act of communication, and self-marketing. In this sense, it allows further valorization of the commodities (see Thesis #10).

Thesis #15: in cognitive biocapitalism, value creation is based pre-eminently on the process of expropriation of the *general intellect* for private accumulation.

The *general intellect* is the outcome of basic social cooperation: it allows the passage from tacit knowledge to codified knowledge as social knowledge. This passage is regulated by the evolution of the juridical forms of intellectual property rights. Such a property is thereby added to that of the means of production, giving private property the control of the process of generation (intellectual property) and diffusion of knowledge (ownership of the means of production). Since the exploitation of the *general intellect* implies the valorization of the very existence of individuals, the process of value creation is no longer limited to the workday but extends to include the entire human existence. This means that the measure of exploitation is not really the time of the workday generating *surplus work* but rather that part of the life span that is necessary to generate tacit knowledge – and hence social knowledge – which is then expropriated by the process of accumulation. The effective and direct forms with which the expropriation of the *general intellect* creates value can be different. Among these, the valorization of commodities through the *branding* process is particularly significant. The value of commodities increases together with the increase of their symbolic meaning and of their ability to create an imaginary which is shared by consumers. Even in this case, surplus value originates from totally immaterial elements created by behavioral conventions and by shared relational activities, just as happens for the financial markets. If private ownership of the means of production implies partly stealing the workday and allowing for the generation of surplus work, private intellectual property is then the theft of social knowledge understood as *common wealth* [*bene comune*]. In cognitive capitalism, creation of value happens through the expropriation of the "common."[24]

Thesis #16: in cognitive biocapitalism, *basic income* is the compensation for work.

The idea of basic income is centered on the concept of "compensation" or "recognition" and not of support or assistance (subsidies, transfer payments, etc.). The logic that justifies its existence is then completely opposed to the doxastic interpretation of the current situation, that is, to measures which would guarantee a continuity of revenue in a temporary, conditioned way.[25] In the present context of cognitive biocapitalism, wealth [*ricchezza*] is divided between those whose life becomes value (all residents regardless of citizenship etc.), on the one hand, and all those (much less) who create value from the private appropriation of common goods [*beni comuni*] (exploitation of intellectual property rights, of the territory, of financial flux, etc.), or who profit from productive and service activities. As a consequence, basic income is by definition *unconditioned* and *perpetual* (for the duration of one's life).

In other words, basic income is nothing other, today, than the equivalent of salary in Fordist times.[26]

Thesis #17: in cognitive biocapitalism, the most adequate structure of welfare is the *commonfare*, or welfare of the common.[27]

The welfare of the common is based on two important concepts. On the one hand, the guarantee of a continuity of unconditioned revenue, disregarding working conditions, professional, social, or citizenship status. This is complementary to any other form of direct revenue, as compensation for the productive social cooperation that forms the basis of value creation, currently expropriated for private rent and profit. On the other hand, access to common material and immaterial goods that allows full participation in social life by way of the free fruition of common natural/environmental goods (water, air, environment) and immaterial common goods (knowledge, mobility, socialization, currency, primary social services).

Thesis #18: in cognitive biocapitalism, the trade unions' keyword "right to work" should be changed into "right to choose work."

We are witnessing an ethical overturning of how we conceive of actual work activity. If in industrial-Fordist capitalism the right to work is the foundation of many national constitutions (the Italian first of all) as well as the first objective of union struggle as a pass to revenue stability and the enjoyment of civil rights, in cognitive biocapitalism, insofar as life itself is productive, the necessity of work has largely taken up a function of blackmailing and control of the actual work activity, and is increasingly less relevant to accumulation. From this point of view, capital tends to reach "autonomy," even though it still depends on the social connections that are inherent in the relationship between labor and capital. In contrast, the right to choose one's work opens the path to autonomous work, and for this reason this objective is not compatible with the current capitalistic valorization or subsumed by it. In other words, if in industrial-Fordist capitalism the right to work was, on the one hand, functional to the process of accumulation, while, on the other, it represented the basic condition for the right to struggle, in cognitive biocapitalism the right to choose one's work is uniquely the right of subversion.

Thesis #19: there is no space for an institutional politics of reform able to reduce the structural instability characterizing cognitive biocapitalism.

No new *new deal* is possible. And this is increasingly true the more we seem to detect measures that favor a re-stabilizing of the process of accumulation. These measures promote a salary regulation based on the proposition of *basic income* and a productive ability founded on free circulation of knowledge. These proposals, from a purely theoretical and economic point of view, could have the effect of exploiting the economies of learning in a better way (a continuity of revenue increments the ability and intensity of learning). They could also better exploit network economies (the free circulation of knowledge augments their diffusion and valorization). As a consequence, they could compensate for the structural instability deriving from the distorted effects of financialization on productive activity and revenue distribution. In any case, these measures would undermine the very nature of the capitalist system, that is to say, the necessity of work, the blackmailing allowed by differences in revenues as an instrument of domination of a class over another, and the principle of private ownership of the means of production (machines yesterday, knowledge today). In other words, we can conclude that in cognitive biocapitalism a possible social Keynesian compromise, one adequate to the characteristics of the new process of accumulation, is possible only in theory, but could not be carried out politically *sic rebus stantibus.* A real reformist politics that can guarantee structural stability of the paradigm of cognitive biocapitalism (which would tend to individuate a form of mediation between labor and capital that is satisfying for both, without paving the way to the overcoming of this very economic system) cannot exist. Let us clarify this point: a possible social compromise based on *basic income* and the free diffusion of knowledge and other common goods undermines the basis, the real foundations of the capitalistic economic system, that is, the necessity of work to live (hence its subaltern condition), and private property as a source of accumulation. Such compromise is not possible, unless it is imposed by force.[28]

Final thesis: as a consequence, insofar as praxis guides theory, in cognitive biocapitalism only conflicts and the ability to create and organize multitude-based movements will allow social progress for mankind.

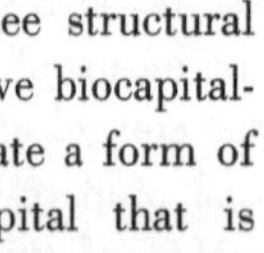

notes

Psychedelic musical support by Grateful Dead, The Phish, Jimi Hendrix.

1 The twenty theses presented here are pre-eminently of a socio-economic nature, and as such they are incomplete. There is no explicit reference, for instance, to the evolution of the structure of ownership (juridical analysis) or the theme of the *common* as a way of overcoming the public/private dichotomy. The very aspect of international relations and the end of the economic hegemony of the United States, with the consequent shift of the economic-financial center to the East (China and India, primarily), is not treated with due detail.

2 As M. Turchetto reminds us:

> The origin of the notion of postfordism does not lie in orthodox Marxism or Workerism. These two currents of thought imported the term and its correspondent definition from France, adapting them to their conceptual apparatus. The *copyright* of postfordism belongs in fact to the French *école de la regulation* . . . (See M. Turchetto, "Fordismo e post fordismo. Qualche dubbio su un'analisi un po' troppo consolidata" in various authors, *Oltre il fordismo. Continuità e trasformazioni nel capitalismo contemporaneo* (Milan: Unicopli, 1999) 1)

One of the first authors to use the term "post-Fordism" was the English geographer A. Amin in his *Post-Fordism: A Reader* (Oxford: Blackwell, 1994). Within the French *école de la regulation*, see B. Jessop, *The Regulation Approach: Governance and Post-Fordism, Economy and Society* (Oxford: Blackwell, 1995); A. Lipietz, "The Post-Fordist World: Labor Relations, International Hierarchy and Global Ecology," *Review of International Political Economy* 4.1 (1997): 1–41; R. Boyer and J.-P. Durand, *L'Après-fordisme* (Paris: Syros, 1998). As far as the Italian debate is concerned, the first text to use the term post-Fordism is S. Bologna and A. Fumagalli, eds., *Il lavoro autonomo di seconda generazione. Scenari del postfordismo in Italia* (Milan: Feltrinelli, 1997). See also E. Rullani and L. Romano, *Il Postfordismo. Idee per il capitalismo prossimo venturo* (Milan: Etas Libri, 1998) and the already quoted critical text by M. Turchetto in various authors, *Oltre il fordismo.*

3 See, among others, T. Ohno, *Toyota Production System: Beyond Large-Scale Production* (New York: Productivity, 1995); G. Bonazzi, *Il tubo di cristallo. Modello giapponese e fabbrica integrata alla Fiat* (Bologna: Il Mulino, 1993); M. Revelli, "Economia e modello sociale nel passaggio tra fordismo e toyotismo" in *Appuntamenti di fine secolo*, eds. P. Ingrao and R. Rossanda (Rome: Manifestolibri, 1995) 161–224; B. Coriat, *Penser à l'invers* (Paris: Bourgois, 1991).

4 See M. Priore and C. Sabel, *The Second Industrial Divide: Possibilities for Prosperity* (New York: Basic, 1984); S. Brusco, *Piccole imprese e distretti industriali* (Turin: Rosenberg, 1989); G. Becattini, *Distretti industriali e sviluppo locale* (Turin: Bollati Boringhieri, 2000). For a critical analysis, see M. Lazzarato, Y. Moulier-Boutang, A. Negri, and G. Santilli, *Des enterprises pas comme les autres* (Paris: Publisud, 1993); A. Fumagalli, "Lavoro e piccola impresa nell'accumulazione flessibile in Italia. Parte I e Parte II," *Altreragioni* 5 and 6 (1996–97).

5 See C. Palloix, *L'economia mondiale e le multinazionali*, 2 vols. (Milan: Jaca, 1979 and 1982); G. Bertin, *Multinationales et propriété industrielle. Le Contrôle de la technologie mondiale* (Paris: PUF, 1985).

6 This term originated in France in the early 2000s from the research of the Laboratoire Isys-Matisse, Maison des Sciences Economiques, Université de Paris I, La Sorbonne, under the direction of B. Paulré, and it is diffused by the journal *Multitudes* with very heterogeneous texts by A. Corsani, M. Lazzarato, Y. Moulier-Boutang, T. Negri, E. Rullani, C. Vercellone and others. On this topic, see also B. Paulré, "De la *New Economy* au capitalisme cognitif," *Multitudes* 2 (2000): 25–42; C. Azais, A. Corsani, and P. Dieuaide, eds., *Vers un capitalisme cognitif* (Paris: l'Harmattan, 2001); Y. Moulier-Boutang, *L'età del capitalismo cognitivo* (Verona: Ombre Corte, 2002); C. Vercellone, ed., *Sommes-nous sortis du capitalisme industriel?* (Paris: La Dispute, 2003); A. Corsani, P. Dieuaide, M. Lazzarato, J.M. Monnier, Y. Moulier-Boutang, B. Paulré, and C. Vercellone, *Le Capitalisme cognitif comme sortie de la crise du capitalisme industriel. Un programme de recherche* (2004). For a more recent analysis, see C. Vercellone, ed., *Capitalismo cognitivo* (Rome: Manifestolibri, 2006); A. Fumagalli, *Bioeconomia e capitalismo cognitivo. Verso un nuovo paradigma di accumulazione* (Rome: Carocci, 2007); Y. Moulier-Boutang, *Le Capitalisme cognitif. Comprendre la nouvelle grande transformation et ses enjeux* (Paris: Editions Amsterdam, 2007). See also the monographic issue "Le Capitalisme cognitif. Apports et perspectives" of the *European Journal of Economic and Social Systems* 20.1 (2007), eds. A. Fumagalli and C. Vercellone, with contributions by A. Arvidsson, L. Cassi, A. Corsani, P. Dieuaide,

S. Lucarelli, J.M. Monier, and B. Paulré, as well as by the editors.

7 See D. Lebert and C. Vercellone, "Il ruolo della conoscenza nella dinamica di lungo periodo del capitalismo: l'ipotesi del capitalismo cognitivo" in *Capitalismo cognitivo* 22.

8 See A. Orléan, *De l'euphorie à la panique. Penser la crise financière* (Paris: Rue d'Ulm, 2009).

9 The terms *bioeconomy* and *biocapitalism* are very recent. The concept of bioeconomy was introduced by A. Fumagalli, in 2004; see "Conoscenza e bioeconomia," *Filosofia e Questioni Pubbliche* IX.1 (2004): 141–61 and "Bioeconomics, Labour Flexibility and Cognitive Work: Why Not Basic Income?" in *Promoting Income Security as a Right: Europe and North America*, ed. G. Standing (London: Anthem, 2005) 337–50, as well as Fumagalli, *Bioeconomia e capitalismo cognitivo*. For an interesting analysis of the concept of bioeconomy, see also F. Chicchi, "Bioeconomia: ambienti e forme della mercificazione del vivente" in *Biopolitica, bioeconomia e processi di soggettivazione*, eds. A. Amendola, L. Bazzicaluppo, F. Chicchi, and A. Tucci (Macerata: Quodlibet, 2008) 143–58 and L. Bazzicaluppo, *Il governo delle vite. Biopolitica ed economia* (Rome and Bari: Laterza, 2006). The term biocapitalism was instead coined by V. Codeluppi, *Il biocapitalismo. Verso lo sfruttamento integrale di corpi, cervelli ed emozioni* (Turin: Bollati Boringhieri, 2008). See also the more recent C. Morini, *Per amore o per forza. Femminilizzazione del lavoro e biopolitiche del corpo* (Verona: Ombre Corte, 2010).

10 See A. Fumagalli and S. Mezzadra, eds., *Crisis in the Global Economy: Financial Markets, Social Struggles and New Political Scenarios* (Cambridge, MA: Semiotext(e), 2010) 237–39. For an in-depth analysis of the evolution of financial markets and the role of the monetary and credit market, see chapter 1 of Fumagalli, *Bioeconomia e capitalismo cognitivo*.

11 For an in-depth analysis of this passage, see M. Amato and L. Faracci, *Fine della finanza* (Rome: Donzelli, 2009) esp. chapters V and VI, 65–90.

12 See S. Lucarelli, "Financialization as Biopower" in *Crisis in the Global Economy* 119–38.

13 See A. Negri and C. Vercellone, "Il rapporto capitale/lavoro nel capitalismo cognitivo" in *Posse* (Nov. 2007); C. Vercellone, "The New Articulation of Wages, Rent and Profit in Cognitive Capitalism," paper presented at Queen Mary, University of London, 29 Feb. 2010. See also C. Marazzi, *The Violence of Financial Capitalism* (Cambridge, MA: Semiotext(e), 2010) esp. chapter 3.

14 For a discussion of the concept of the "common," see M. Hardt and T. Negri, *Commonwealth* (Cambridge, MA: Harvard UP, 2009).

15 See R. Nelson and S. Winter, *An Evolutionary Theory of Technical Change* (Cambridge, MA: Belknap, 1982).

16 This argument can be presented in terms of *tacit* and *codified knowledge*; see F. Malerba and L. Orsenigo, "Knowledge, Innovative Activities and Industrial Evolution," *Industrial and Corporate Change* XII (2000): 289–314.

17 See E.M. Mouhoud, "Global Geography of Post-Fordism Knowledge and Polarisation" in *The Hardship of Nations*, eds. B. Coriat, P. Petit, and G. Schméder (Cheltenham and Northampton: Elgar, 2006) 300.

18 See C. Marazzi, "Capitalismo digitale e modello antropogenetico del lavoro. L'ammortamento del corpo-macchina" in *Reinventare il lavoro*, eds. J.L. Laville, C. Marazzi, M. La Rosa, and F. Chicchi (Rome: Sapere 2000, 2005) 107–26. Here is the complete quotation that defines the concept of the anthropogenetic model of production:

> A model of production of man through man, in which the possibility of cumulative and endogenous growth is due, above all, to the development of the education sector (investment in human capital), the health sector (demographic evolution, biotechnologies) and the cultural sector (innovation, communication, creativity). (Ibid. 109)

19 J. Halloway writes the following:

> The center of class struggle is located here: it is a struggle between creative action and abstract labor. In the past, we always thought of class struggle as a struggle between labor and capital, thus understanding labor as abstract, wage-earning labor. As a consequence, the working class was defined as the class of wage-earners. This is wrong. Wage-earning labor and capital are

mutually completing, the former being a stage of the latter. Doubtlessly, there is a conflict between wage-earning labor and capital, but it is rather superficial: a conflict on salary levels, on work conditions, on the length of the work day. All these things are important, but they presuppose the existence of capital. The real threat to capital does not come from abstract labor, but from useful labor or creative action, because it is the latter that is radically opposed to capital, that is, to its own abstraction. Creative action says "No, we will not let capital control us; we need to do what we think is necessary or desirable." (See J. Halloway, "Noi siamo la crisi del lavoro astratto," intervention at the UniNomade seminar, Bologna, 11–12 Mar. 2006, manuscript)

20 See A. Arvidsson, *La marca nell'economia dell'informazione. Per una teoria dei brand* (Milan: Angeli, 2010).

21 See A. Fumagalli and C. Morini, "La vita messa a lavoro: verso una teoria del valore-vita. Il caso del lavoro affettivo," *Sociologia del lavoro* 115.3 (2009): 94–116 (forthcoming in English as "Life Put to Work: Towards a Life Theory of Value" in *Ephemera*).

22 See Morini, *Per amore o per forza*.

23 See ibid.

24 I cannot develop here an in-depth analysis of the theme of the "common." On this topic, see M. Hardt and T. Negri, *Empire* (Cambridge, MA: Harvard UP, 2000), and Hardt and Negri, *Commonwealth*.

25 Such as, for instance, the French RMI and analogous apparatuses, which simply function as social shock absorbers and promote the return to work.

26 See A. Fumagalli, "Per una nuova interpretazione della teoria del *basic income*" in *Reddito per tutti. Un'utopia concreta per l'era globale*, ed. Basic Income Network (Rome: Manifestolibri, 2009) 125–40.

27 See A. Fumagalli, "Trasformazione del lavoro e trasformazioni del welfare: precarietà e welfare del comune (*commonfare*) in Europa" in *L'economia della precarietà*, eds. P. Leon and R. Realfonso (Rome: Manifestolibri, 2008) 159–74. See also chapter 9 of Fumagalli, *Bioeconomia e capitalismo cognitivo*.

28 For further research on this topic, see A. Fumagalli and T. Negri, "John Maynard Keynes, Capitalismo Cognitivo, Basic Income, No Copyright: è possibile un nuovo 'New Deal'?" in *Quaderni di Ricerca del Dipartimento di Economia Politica e Metodi Quantitativi*, Università di Pavia, no. 211 (2008). See also Thesis #9 in "Nulla sarà come prima. Dieci tesi sulla crisi finanziaria" in *Crisi dell'economia globale. Mercati finanziari, lotte sociali, e nuovi scenari politici*, eds. A. Fumagalli and S. Mezzadra (Verona: Ombre Corte, 2009) 222–26.

DYSLEXIA AND THE ECONOMY

christian marazzi

translated by sabrina ovan

In May 2002, the economics and financial magazine *Fortune*, known for its yearly list of the world's "500 largest companies," published an article on dyslexia, the "linguistic disorder" common to a great number of businesspeople over recent years. It is estimated that about 20 percent of the population of the United States might be affected by some degree of dyslexia. This percentage is particularly over-represented among executives. Months after the burst of the first great crisis in the economy of knowledge, this "corporate anomaly" has become a stimulating field of research, one more engaging than the delinquent normativity of the executives of Enron, WorldCom, and of all those companies systematically using scams and lies in order to secure success in the 1990s. Confronted with the crisis of the delegitimized executives, corrupted by the financial world at the expense of innovation, emergent capitalism needed an anthropologic shock treatment in order to define the profile of the new Schumpeterian entrepreneur.[1]

The personal and professional histories of *dyslexic achievers*, the successful dyslexics interviewed by *Fortune*, reveal that this specific learning disability (SLD) is in fact a *virtue*, a talent that schools and institutions – the dominant language system – are unable to understand and value. The personal histories of dyslexic managers allow them, on the other hand, to maintain that their professional success is not due to nominally effective processes of therapeutic normalization (that is, *in spite* of their SLD) but to the fact that their "gift" could be put to use *thanks to* the specific nature and functioning of the new economy. What was considered a linguistic handicap and a pathology less than a generation ago is potentially a "competitive advantage" for digital capitalism.

language nomads

According to Ronald Davis, sculptor, engineer, founder of the Reading Research Council Dyslexia Correction Centre, and author of *The Gift of Dyslexia*, the mental faculty causing dyslexia is literally a gift, a natural skill, a "perceptual talent." When it is not destroyed or prematurely repressed by parents or the school system, this talent will develop into extraordinary creativity, and into a type of intelligence that is superior to the norm. Not all dyslexics will develop the same talent, and every case of dyslexia is different, but there are some fundamental, shared abilities: the ability to alter and create perceptions, an extreme awareness of the environment in which one is immersed, above-average curiosity, the ability to think through images, intuition, introspection, multi-dimensional thought and perception, the ability to feel thought as a reality, a vivid imagination. These qualities are shared by all managers interviewed by *Fortune*.

To Paul Orfalea, every sentence has always been worse than an Egyptian hieroglyph. At school, he developed a symbiotic relationship with his classmates, to whom he commended all written homework, while he concentrated his full-time efforts on the copy-machine. This is how he started to conceive of Kinko, the outsource service for companies, offering print jobs on any kind of material.

Losing oneself within a text, and moving physically inside the school institution as if on a large chessboard, would develop for the young Charles Schwab – the future inventor of *discount brokerage* – a strong tactical sense, a flexible mind and a specific risk-taking ability. For Bill Dreyer, inventor of the first automated protein sequencer, being a dyslexic is equivalent to having a built-in CAD [computer-aided design] in the brain, allowing the business world to be treated as a tri-dimensional space. John Chambers, founder of Cisco, the iconic *new economy* firm producing network communication systems, attributes his success to the ability to conceptually "jump" from a point to another:

> I imagine a game of chess on a dimensional cycle with different layers, and I can almost play it outside of my mind. But it's not a game. It's business. I never do one move at a time. I can usually predict the potential result and the place where Y-s will be found along the way.

Sally and Bennett Shaywitz, directors of the Yale Center for Learning and Attention, maintain that the difficulty in decoding and focusing on *phonemes* develops, in dyslexic patients, the ability to rapidly see or perceive the big picture, the *context* within which they need to operate: "They have no choice, it is a question of survival." Unlike the verbal conceptualization characterizing the usual way of thinking through the *sounds* of words, dyslexics think mainly through mental *images* of concepts and ideas. The internal monologue of the dyslexic is scarce or non-existent; he does not internally hear what he reads, unless he reads out loud. During his reading process, this type of internal "deafness" is substituted by "talking images" which function like vehicles for the meaning of words themselves.[2]

The analysis of dyslexia, then, underlines a more general problem, one well known to semioticians and research groups that study the mind, that is to say, the evaluation of the *cognitive impact* of new information and communication technologies. More specifically, research stresses the question as to whether there is a structural compatibility or a heterogeneity between word-based and image-based cognition. Dyslexic entrepreneurship, as we shall see, seems to resolve this diatribe by favoring the first hypothesis.

Verbal thought is *linear* through time, in the sense that it follows, phonically rather than graphically, the consequential structure of language.[3] When we use verbal thought, we mentally compose sentences, one word at a time. Verbal thought develops at the speed of language, on an average of 150 words per minute, or 2.5 per second. Non-verbal thought, on the other hand, is *evolutional*. The image "develops" while thought adds new concepts. For this reason, non-verbal thought is much faster than linear-verbal thought, and this is surely an advantage for entrepreneurial activity in a competitive setting. It is estimated that image-based thought consistently develops at a speed of about thirty-two images per second, or a frequency of 1/32 of a second, the same speed as visual perception.[4] The speed of image-based thought (or "glance" intelligence, as we may call it) allows us to explain the ability of *intuition* among dyslexics, in the sense that they become aware of the *product* of thought processing at the moment in which it takes place, but they are not aware of the production *process* of this same product. A dyslexic, we could say, knows the answer without knowing why he knows it.

It is important to point out that everyone thinks in *both* ways – verbal and non-verbal – but we also all have the tendency to specialize, to practically use one mode as the dominant variety of thought and the other as secondary. It is worth mentioning, here, Franco Lo Piparo's critical reflection on those who, facing the diffusion of communication technologies and their iconic-visual representation, create a strict dichotomy/competition between verbal and visual language:[5]

> We should resolve to do away with the idea that verbal language is a development of gestures and think of the birth of language as a phonic-visual event (one that sees gesturing, graphics and articulate voice as cooperating agents), rather than simply as a phonic-auditory event. The claim that the first words were accompanied by forms of graphic representation, together with gesturing which possessed a communicative aim, is the only one compatible with the reciprocal implications of the four constitutive elements of the human animal [erect position, transformation of the upper limbs into hands, use of tools, language[6]].

If language is then born multi-sensorial, or, to use contemporary vocabulary, multi-medial, the consequent passage from a Gutenbergian typographic environment to a virtual one does *not* require a physiological revolution, only different cooperation between verbal and visual language. "The new communication technologies," Lo Piparo continues, "do not produce any new sensorial order: rather than innovating the hierarchy of senses, they reproduce, in a more complex form, a sensorial co-order that is as old as mankind." He then concludes:

> Information technologies clearly constitute a revolution, not because they modify the order of senses (Simone), nor because they help images in their competitive contest with language (Parisi). In these fields, new technologies do what every other technology does: they conserve, transform and complicate what is already in nature. Their revolutionary charge must be found elsewhere.

Dyslexia as a "specific learning disability" starts to manifest in the school years, when the child is first exposed to verbal language. If in pre-school age the child has developed primarily non-verbal thought, the impact of having to learn how to write can indeed be traumatic. This congenital difficulty will be evidenced by the reading of words that cannot be pictorially represented in their meaning. In non-verbal thought it is not difficult to think of the word *elephant*, knowing what an elephant looks like. The animal we call "elephant" is the literal meaning of the word elephant: to see its image is to see its meaning.[7] The same point can be made for the verbs to fly, to sleep, to watch, because we all have seen, felt or heard the actions these words describe. We necessarily need to keep in mind the correlation, *even* in the case of dyslexics, between the ability to perceive an image and the ability to use language: "Could the image of an angel exist," Lo Piparo asks, "if there were no word 'angel,' with the meanings associated with it, or phrases such as 'this is an angel, this is not an angel'?" This means that it is not necessary to have a direct experience of something to know what it is (we do not need to visit Africa to know what an elephant is), but rather it means that even in the case in which non-verbal thought prevails, as in dyslexia patients, verbal language has somehow achieved its aim *regardless* (for instance, when parents tell a virtually dyslexic child "*this* is an elephant, this is *not* an elephant").[8]

Those who think non-verbally cannot think of those words whose meanings cannot be represented visually. Knowing the way an "*a*" looks does not mean we can think of an "*a*." To see the letters T-H-E forming the word *the* does not mean we can see the word's meaning.[9] There are about 250 words which create problems for the majority of dyslexics. They are part of spoken vocabulary, but dyslexics cannot form a mental image of their meaning.[10] Davis calls them "trigger words" because, having abstract meanings, often a number of different meanings, they cause *disorientation*, the feeling of confusion dyslexics feel while reading a text. Whenever a dyslexic patient encounters one or more trigger words during his reading, the evolutional development of the mental image forming with the sentence comes to a stop. It is like seeing a sequence of disconnected images with white spaces in between them. The disorientation caused by the encounter/clash with trigger words means that symbolic perception is altered and distorted to the point that it becomes very difficult, if not impossible, to read or write.

Trigger words strongly alter the logical functioning of *syntax*, the part of grammar formulating the laws of sentence composition, those which concern the functions of words and their organization into series called propositions. Because of a deficit in logical-syntactical operations in the dyslexic patient, the first perceived meaning of a word, the one that comes up first in relation to the word (the "elephant" we were discussing above) immediately inveigles thought; the latter cannot take the necessary distance from it, it cannot go forward, either vertically or horizontally, it cannot change meaning and thus complete the understanding of the proposition. It is as if every word were perceived "in itself," immediately, out of context and out of syntax. As André Rey pointed out in *Monographies de psychologie clinique*, "the subject cannot rapidly get rid of the literal elements perceived – sometimes only of a few – and skip to other connections. He has to set loose, and in order to do so he skips referents, resulting in disorientation and incoherent repetitions."

The interesting aspect is that this disorientation, this displacement of perception, is the same mechanism that dyslexics develop and use to recognize objects and real facts of life in their environment, *before* they learn to read. From a very young age, dyslexics unconsciously use disorientation to acquire multi-dimensional perception. By displacing their senses, they can have multiple visions of the world. That is to say, they perceive things from more than one perspective and they retain, from such perspectives, more information than other people do. Since they do not work with verbal-linear logic, dyslexics *globalize* starting from an "*embracing perception*," and so often guess, juxtapose, repeat, anticipate. It is not difficult to recognize, in this mode of thought, a form of entrepreneurial virtue. This is especially evident in a globalized world, where a multiplicity of heterogeneous and disorderly factors, with multiple and often contradictory movements, concur in defining the economic context in which we operate.

It should be noted that disorientation is a common phenomenon in everyone, a natural function of the brain. It usually appears when we are overwhelmed by thoughts or stimuli. It also appears when the brain receives information from different sensorial organs and tries to link them together. The *active* use of disorientation in the case of dyslexics in pre-school age is, to this day, a controversial topic. According to Glen Rosen, Associate Professor of Neurology at Boston's Beth Israel Deaconess Medical Center, in the dyslexic's brain, dull clouds of neurons deviate from their regular path (probably during fetal development) to form small groups called ectopias (which literally means *out of space*). These nomadic neurons cause a flow of connective differences in the cerebral wiring. Ectopias obstruct certain nerve fibers, which are not able to go where they should, and so they migrate in random directions, connecting regions of the brain that are not usually connected. Further analyses of cerebral activity via magnetic resonance apparently show that dyslexics have great difficulty reordering phonemes, because sound stimulates the anterior part of their brain. For non-dyslexic readers, on the other hand, the posterior part of the brain is activated by the sound of a phoneme. Following this claim, an Italian-British essay published in *Science* has found that the seriousness of the symptoms derives from the complexity of the native language's spelling. Even though the neurological basis of dyslexia is universal, the chances of suffering from dyslexia will be lower for Italian speakers, owing to the more regular spelling of the language, than for English speakers.

I do not have the expertise to express an opinion on the scientific foundation of these theories. I think I can say, however, that dyslexia is assuming the characteristic of a *social* phenomenon (so much so that it is becoming a *managing skill*). Such a phenomenon is probably induced by the diffusion of multi-media and communication technologies (TVs, computers, video games), and the cognitive impact these have on children from the first months of their lives. If this impact, as a social phenomenon, puts "the faculties of thought and language *in themselves*" at the center of our analysis, then, possibly, the study of dyslexia will confront the *dilemmas* and the *conflicts* that run through the field of research on human technoscience mentioned in Massimo De Carolis's *La vita nell'epoca della sua riproducibilità tecnica* [*Life in the Age of its Technological Reproduction*]. "The basic problem," De Carolis writes, "is the relation among human faculties, of which every practical skill is a particular variation, and the biological constitution where these faculties take shape."[11] Our earlier considerations on the phonetic-visual origin of language and on the multi-medial origin of the human animal, that is to say, on the "genetic" cooperation between verbal and visual language, will allow us to consider the specific talents of dyslexics as constitutive of the more general faculty of human language. Under this perspective, neurobiological research dealing with dyslexia runs the risk of taking into consideration and subsequently analyzing dyslexia in a *reductive* way, trying to isolate, that is, its neurologic cause in order to *optimize* therapy. Possibly, inventing a pill that gives linearity to visual thought in the dyslexic patient.

If, instead, we look at the *practical* experience of reading and writing teachers, tutors who teach dyslexics to study in a normal way, we will understand that the essential objective is to allow dyslexics to acquire *technical* skills that would let them perceive symbols in a bi-dimensional way. These skills would help to overcome the aspect of non-alphabetic vision (multi-dimensional, visual thought) that makes the reading of linear-consequential written texts very difficult. Teaching models for reading and writing are based primarily on the *active* use of disorientation. Disorientation in dyslexics is linked to the "mental epicenter of perception," the *mind's eye*, which is completely mobile in dyslexic patients.

> When dyslexic people look at an alphabet letter and disorient, within a split second they see dozens of different views – from the top, the sides, and the back of the letter. In other words, the mind's eye is mentally circling around the letter as though it were an object in three-dimensional space. It's like a helicopter buzzing around, doing surveillance on a building. This is the disorientation function hard at work, trying to recognize the object.[12]

If the object is a concrete image, the disorientation induced by the mobility of the mind's eye constitutes a true "perceptual talent." If the word is not immediately translatable into a stable iconic meaning, as in the case of "trigger words," disorientation leads to confusion and growing stress. Letters and words literally slide outside the text, they move, they are inverted and for these reasons reading becomes a strenuous task.

We need to note that the *positive* function of disorientation and of the *mobility* of the mind's eye is a *recent achievement* in the field of research on dyslexia. We will have the chance to approach this question again below, but it is important to point out that until the end of the 1960s, a typical effect of disorientation, specifically the *absorption in what is perceived and concrete* that blocks the reference to the organizational principles (syntax) in dyslexic visual thought, was frowned upon. This effect was considered as a *disease* to be cured. As long as one continued to think hierarchically of the distinction between analytic-synthetic intelligence (typical of verbal-linear thought) and syncretic-analogical intelligence – "a prisoner of the perceived concreteness" – (typical of visual thought), dyslexics' intelligence was considered "virtual, inefficient, unproductive." It was a sort of "*unarmed intelligence*, fighting a powerless battle against confusion in search for an orientation which will remain fluid and uncertain."[13]

I believe we can affirm that the basis for the distinction between these two types of intelligence was the Cartesian dualism between body and mind, following the claim that *detachment*, the *subtraction* of reflection from the charge of the forms of perception, is a fundamental element in the constitution of analytical-synthetic intelligence.[14] This is a "historically determined" dualism that reflected the separation or functional juxtaposition of production and communication, typical of the Fordist period. At that time, in a society that was already strongly literate, and where advertising was the only form of communication that did not rely on the written word, the elimination of dyslexia was the primary task of doctors, psychologists and teachers. The "linguistic turn" inscribed in the economy by the transition to post-Fordism and by the diffusion of multi-media technology – the becoming-communication of every productive act – has certainly redefined the basis on which we base our understanding and our theories of dyslexia.

Even though the idea of a moving mind's eye might seem bizarre[15] (it is not to dyslexics), it is empirically proven that a dyslexic can learn to read, write and study from the moment in which he is able to locate the point of orientation blocking the distorted perceptions of dyslexia.[16]

> The Davis Orientation Counseling procedure [. . .] is used to teach the person how to control the position of the mind's eye and move it to the optimum viewpoint for real world perception, especially for reading. The goal of Orientation Counseling is not to stop the person from disorienting, for disorientation is a valuable talent. Orientation Counseling trains the person to turn disorientations on and off at will.[17]

This is not a therapy or a cure, but rather a *pedagogic tutoring model* aimed at the control of orientation and stress-reducing techniques, in order to infuse confidence with respect to letters, and improve the sense of time passing.[18] In other words, dyslexia is recognized as a skill among others within the more general human faculty of language. As such, dyslexia is usually manifested as an entrepreneurial ability.

dyslexic economy

Since the end of the 1960s the interpretation of dyslexia has changed radically. After being called the "disease of the century," it has now become a gift, a precious talent to protect and valorize through valid tutoring activities. We cannot affirm, of course, that dyslexia is sufficiently known in the school system today. The dyslexic's impact with learning verbal thought is still, in the majority of cases, a trauma and a cause for humiliation and exclusion. The fact remains that in the last thirty years not only did the evaluation of dyslexia take a dramatic turn but dyslexia grew and manifested itself to the point of becoming a social phenomenon.[19] A hypothesis worth exploring is that this phenomenon can be interpreted with reference to the cognitive impact of the third industrial revolution. Dyslexia as an "entrepreneurial skill" can be explained through the nature of the digital economy itself.[20] This is what I will try to do, while I outline a few characteristics of new capitalism and their relation to dyslexic behavior.

digitalization and social cooperation

One of the consequences of the diffusion of information and communication technology is the transformation of the division of labor as an effect of *technological delegation*. After an epoch of delegation to specialists, typical of the division of labor in the Fordist industrial economy (delegating to the doctor, to the specialized worker, to the lawyer, etc.), information technology allows us to delegate to machines a series of occupations that are usually part of writing practices and oral communication: elaboration, transmission and storage of data. Machines are also in charge of certain decisions, those which amount to fundamental and basic connections in linear language and writing. The availability of complex multi-medial machines that do not simplify or reduce the complexity of the world by means of linearization has important consequences for the economy [*l'agire economico*]. On the one hand, social practices of intertextual reading and writing (engrained in multimediality) modify the way we organize texts, treat information, and define the horizon of what is possible. All definitions of the hypertext underline its character of *non-consequential* writing, its being a structure that immediately opens the opportunity for non-linear reading and thought. On the other hand, the diffusion of new multimedial technologies achieves a multi-sensorial immersion of the body in its environment, its being in sync with the world. More than a return to the immediacy of the primitive man, it is the *biological* nature of language to emerge strongly from this reordering of verbal and iconic vision, in particular, its being a "vital activity of natural organs," of *all* the organs of the human body.

In an environment filled with multi-medial technologies, where all bodily senses are activated and modulated in order to deal with the disorientation caused by non-consequential perceptions, not only do dyslexics feel more at home, they develop aspects of its way of functioning which become authentic professional abilities. In dyslexics, multi-medial thought, the ability to experience thought as if it were reality and use simulation as a laboratory for creativity[21] are *necessarily* put to work *through social cooperation*. More and more dyslexics are becoming CEOs and entrepreneurs due to their intuition of "seeing" the correct strategy to make the collective labor force cooperate and mobilize.

> They invest heavily in the latest dictation and video equipment – anything that transmits information in a form other than writing. They will rely on trusted subordinates to read things for them and relay messages that must be delivered in writing.[22]

Cooperation transforms dyslexic illiteracy into *social* productive force. This transformation, however, is mainly possible owing to the proliferation and availability of multi-media technologies.

Social cooperation works as a trigger for abstraction along the productive process. It organizes the passage from the concrete/perceived to the "economic grammar" of the process of value production. For the dyslexic entrepreneur this means being able to organize the transition from the syncretic-analogical plane (his own) to the analytic-synthetic one (that of the cooperating workforce), the passage from creative praxis to *profit* production. If, in the pre-logic and syncretic universe of dyslexics, feelings have the double characteristic of being absolute and self-centered, while the advent of analytic reasoning, of generality and relativity to relations (that is, the access to the verbal-analytic universe) coincides with a progress of the *affects* (the ability of placing oneself in rapport with others and of conceiving sentimental relations among others), then the economic success of the dyslexic entrepreneur (based on possessive individualism) leads us to question whether affectivity and ethical consciousness are *exclusive* characteristics of social cooperation. For the dyslexic *entrepreneur*, the logic of profit is that which *decides* when confronted with the ambiguity of values as such, as the latter are manifestations of a creative activity that is an end in itself (non-instrumental, without an aim). On the one hand, the entrepreneur's dyslexia puts the productive force of social cooperation to work; on the other, he would not be an entrepreneur if he were not dyslexic to a certain degree.[23]

growing profit and evolutional thought

With the abolition of the border between goods and services, and the emergence of an immaterial economy centered on the relation between man, ideas and images, the content of information and the interaction of products intensify, altering the nature of commodity-products. The economic value lies less in the commodities' physical properties and always more in their ability to allow access to immaterial tasks.[24]

As Jeremy Rifkin affirms, in the world of neo-capitalism, access usually takes the place of property,[25] renting takes the place of buying. The hegemony of services modifies the very concept of consumer goods. Products are no longer thought as having immutable characteristics with a definite value, but as *evolutional products*, susceptible to all potential modifications and capable of offering services with an added value. The material product functions mainly as a *support* to the distribution of services, one that allows a long-term relationship between the company and the client (through the personalization of customer service). The evolutional character of these products also acts retroactively upon the organization of the companies: the networking organization prevails, together with the "connectivity" of companies among each other and between the company and the clients; the horizontal structure of the productive process changes the companies' organization charts. The new enterprise itself becomes evolutional, that is to say, it is forced to "breathe with the market" (modify production to the oscillations in demand) and to be organized according to the logic of *cognitive distribution of labor* that valorizes intellectual capital. The centrality of immaterial goods, that is to say, knowledge and cognitive work, displaces the productive force of scientific innovation (Marx's *general intellect*) from the machines to the living bodies of the workforce. As a result, *cumulative technologies* tend to prevail over *discrete technologies*. In emerging sectors such as the advanced engineering industry, electronics, and biotechnology, the development of technologies tends to proceed through the continuous placing on the market of new products that are usually only slightly more advanced than the preceding ones. This happens independently of costs and expectations of financial revenue.[26] Scientific and economic development presents an evolutional and cumulative character in and of itself.

There is an isomorphism between the continual flux of incremental innovation and the evolutional thought of dyslexics. Speed and the combinatory art of visual thought, the fact that the image grows according to the mental process's cumulative additions of sub-concepts to a comprehensive concept in a non-consequential linguistic game, on the one hand evoke the modality of Schumpeterian entrepreneurial agency, but on the other allow its limits to be overcome, in particular the economic distinction between innovation and invention.[27] In Schumpeter, an innovation that destroys the routine of the economic cycle is the result of a different (innovative) combination of the same productive factors carried out by the innovative entrepreneur. Innovations start a productive leap of the economic system as a whole, destroying its normal functioning (the routine of the business cycle).[28] This typical characteristic of evolutional thought combines the constitutive parts of a text in an innovative and different way, aggregating them in an incremental series that does not follow the rules of syntax and the grammar that assigns a meaning on the basis of logical-linear principles. At the same time, evolutional thought, with its typical multi-dimensional character, internalizes the *invention* within the very economy of innovation. Even though Schumpeter distinguished invention from innovation, considering the former outside of the economic field, every form of invention permeates the field of the economic application of innovation. The sources of economic-industrial innovation are not only economic. On the contrary, as new technologies pervade the whole of society and the recreational competences [*competenze extra-lavorative*] of the labor force are put to work, the sources of innovation multiply and extend to cover the most diverse areas of social life.[29] New capitalism functions logically through its way of interiorizing *of necessity* what is outside the specifically economic field, through its capacity to transform all changes, including infinitesimal and incremental inventions, into innovative leaps.

the fragility of power

The dyslexic dimension of the new economy finds its most complete expression in the processes of *financialization* that accompanied the digital revolution of the 1990s. The centrality of the stock market in financing economic growth and in the launching of new Internet companies; the hijacking of collective savings (retirement funds and mutual investment funds) to stocks; the chronic uncertainty of investors; the speculative bubble and its burst in the first months of 2000, all of these processes reinstated in economic analysis the notion of *mimetic behavior*, elaborated by Keynes in the 1930s. It is not just natural to imitate other people's behavior when in doubt about a course of action, that is, if we have a *structural deficit* of information: in financial markets, it is also *rational*.[30] This overturns the neo-classic theory of rational anticipation, according to which economic agents decide in isolation from each other, using at best the information they have to form anticipations, as long as the "true" model of the economy is known. Mimesis, on the other hand, defines the decisions/choices of a multitude of investors as a gregarious-collective process that transcends individual beliefs (based on the presumed rationality of the neo-classic *homo oeconomicus*) and encounters its full rationality within so-called *conventions*. Conventions are "collective beliefs," "common opinions" born *within* the ensemble of participants in the market, which end up imposing themselves as *cognitive constrictions* on the participating individual subjects. The underlying logic of conventions implies that we should take into account *rumors*, as they are something that circulates, and as such have an impact on all other operating factors. Whether or not they are true has less importance in the short run than their real and immediate effect on bond prices, which have been elected by convention.

In the financial sphere, investors' predictions can be focused on a title or another for reasons that are completely independent from fundamental underlying values. Think, for instance, of the role played by symbolic numbers in stock-market dynamics, as in the case of the 10,000 mark in the Dow Jones. Since these numbers have an intrinsic saliency, they are chosen by a multitude of investors and they become "commonplace," a convention. The "Internet convention" of the late 1990s or the more recent "China convention" are also market tendencies born within a community of investors following the logic of self-referential rationality: the first referent from which the agents will derive their behavior is not a norm or a value outside of the mimetic process but a norm or a value produced, in the end, by the mimetic process itself, by the opinion of the majority. From this point of view, self-referentiality must be interpreted as an extraordinary mechanism of the amplification of rumors.

Conventions are historically determined, in the sense that they regroup a large number of *heterogeneous* factors which concur in determining the qualitative tendencies of the market. The Internet convention, for instance, synthesized the diffusion of communication and information technologies at a domestic level, the impression of an unlimited development of the digital market, the infinite power of interconnectivity, and the social aspect of online knowledge. In its intersection with finance, this very convention evoked the possibility of obtaining revenues detached from labor and the life–work cycle (pension-income).[31] We could affirm that in the process of creation-election of a convention, the syncretic-analogical prevails over the analytic-sequential.

> This depends on imagination rather than logic, on poetry and mood rather than mathematics. It is interesting to note that traditional game theory has never been interested in this question: that people can coordinate themselves in this way is something that we cannot comprehend by reasoning *a priori*.[32]

The flow of the stock market index in which conventions are condensed reflects the multi-sensorial and multi-dimensional dynamics that are in conflict with the linear-sequential logic forming the basis of the economic circuit (the Marxian money–commodities–money series). The language of the stock market *exceeds* the language of economics, the quantitative measures of offer and demand, of savings and investments, and their linear relations. Financial crises simply reveal the fragility of the power of financial markets' "dyslexic" acting, the contradictions between the syncretic-analogical agency of the subjects that compose the market and the linear-consequential logic of the economy.

The analysis of conventions demonstrates the existence of "bad" conventions,[33] which were chosen due to specific historical events. The "Internet fever," just like the tulip fever in the seventeenth century, may be considered a posteriori as a "collective delirium." At any rate, the collective beliefs surrounding the conventions have an undeniable *creative* role. What a multitude of agents *think* and the way in which they *represent* the world influence prices, hence the relations among economic agents. The main obstacle to emerging from a crisis, Keynes wrote, is not the objective scarcity of capital but the way in which individuals represent for themselves the normal value of the interest rate. The barrier between men and their happiness is not some natural exogenous constriction, only the belief in it. This happens because conventions are not anchored to a secure, definite knowledge, but are the result of collective processes that are intrinsically evolutional and thus can create their own syntax, the grammar of their happiness and unhappiness.

In dyslexic capitalism, the creative power of the actions of mankind frees itself from the constrictions imposed by the linear logic of the market economy. A crisis reveals its internal becoming, the "delirious" alternation of multi-sensorial creativity and disciplinary economic order.

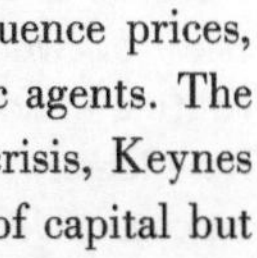

notes

1 This text takes the difference between entrepreneur and manager for granted. This difference arose in the mid-twentieth century with the development of corporations and the administrative changes in large companies analyzed by Berle and Means (1932), that is to say, with the birth of executive capitalism and after the disappearance of the classic entrepreneur (see Robin Marris, *Managerial Theories of the Firm*). This distinction is close to the development of the stock market and the separation between management and property, as well as to the analysis of the nature of innovation; as Joseph Schumpeter had already underlined in 1942:

> The [innovative] function [of the entrepreneur] is already losing importance and is bound to lose it at an accelerating rate in the future [...] Innovation itself is being reduced to routine. Technological progress is increasingly becoming the business of teams of trained specialists who turn out what is required and make it work in predictable and controllable ways. (*Capitalism, Socialism and Democracy* (London: Routledge, 2003) 132)

However, as I will try to explain in this article with reference to the dyslexic entrepreneur, the difference between "destructive innovation" and "routine innovation" (see William Baumol, *The Free-Market Innovation Machine* (Princeton: Princeton UP, 2002)) is no longer so clear, especially after the new innovations of the 1990s concerning the general application of new communication technologies, of networking production, and, above all, of diffuse knowledge. See, for instance, Michael J. Mandel's book *Rational Exuberance: Silencing the Enemies of Growth* (New York: HarperCollins, 2004), in which a *Business Week* economy reporter defends the centrality of *breakthrough innovations* (such as the combustion engine or the Internet) in order to promote consistent, long-term growth. Within the same question, a special mention should be made of Maurizio Lazzarato's study on the economic psychology of Gabriel Tarde (*Puissances de l'invention* (Paris: Seuil, 2002)).

2 This, we shall see, allows us to inscribe the dyslexic's absence of internal monologue within the linguistic skills that universally distinguish man as *speaking animal*.

3 Giuseppe Longo, for instance, writes:

> We have always been trained to speak in a linear way; it is a physiological fact; for millennia we have also been used to writing in a linear way. This means that all our culture rests on the hinges of linearity, so much so that even scientists, who think they are less ingenuous than other people, describe the world in a linear manner. For instance, they adopt a consequential description and superimpose it onto a causal description: A happens, then B happens, then C happens. From here they will say: A happens, then, as a consequence, B happens, then, as a consequence, C happens etc. In this way, the succession becomes a chain of causes and effects. ("L'avventura complicata delle nuove tecnologie," *Cenobio* LI (Jan.–Mar. 2002))

4 Note that if a stimulus is present for less than 1/25 of a second, but more than 1/36, we enter the *subliminal* sphere. The brain receives the stimulus, but the person is not aware of what they received. If the stimulus is a part of a continuum, then it is confused with the images that precede and follow. Dyslexic image-based thought is then faster than the incidence of awareness (1/25 of a second), but slower than the subliminal limit.

5 The reference comes from a review of Raffaele Simone's *La terza fase. Forme di sapere che stiamo perdendo* (Rome and Bari: Laterza, 2000) and Domenico Parisi's, *Scuol@.it. Come il computer cambierà il modo di studiare dei nostri figli* (Milan: Mondadori, 2000). "Both authors are convinced that the computer revolution will strongly modify human cognitive systems. The difference between them is that, while Simone underlines the regressive and negative aspects of the phenomenon, Parisi emphasizes the liberating and positive aspects" (see Franco Lo Piparo, "I sensi, le immagini, il linguaggio, e la rivoluzione conservatrice dell'informatica," *Sistemi intelligenti* XII.3 (2000)).

6 The erect position, Lo Piparo writes, frees the arms from locomotive functions, and transforms them into hands. The hand, in its turn, is one of the causes of a series of fundamental events: (i) it frees the mouth from attack/defense functions, favoring the acquisition of an articulated voice; (ii) it is particularly apt for the creation of tools; and (iii) it can point at, indicate, and trace signs and images. To confirm the natural cooperation of iconic and auditory vision, the author refers to

sign language as an example: "The hand and the phono-articulatory apparatus, being neuro-genetically connected and genetically cooperative (not competitive), are also able to exchange functions when one of the organs is absent or in deficit." We cannot derive the same sensorial cooperation, at least in sign language, from Simone's approach:

> Alphabetic vision must be accurately distinct from other sensorial modes. It has, in fact, shared characteristics with non-alphabetic vision and also with hearing, but is more similar to the latter. Like non-alphabetic vision, it is obviously mediated by the eye, but just like hearing, it adds to this trait that of being linear, that is to say it can be applied to objects in a series. (Simone, *La terza fase* 20 n. 21)

In reality, sign language presents an ability in terms of linearization that is equal to that of people with normal hearing ability.

7 We could state that the image of the elephant is *inseparable* from the semantics of the word elephant, except that for dyslexics the degree of iconicity that is mentally evoked by the word prevails over its degree of arbitrariness.

8 Wittgenstein's theory of language as a mirror, used by Lo Piparo to introduce the book by Antonino Pennisi (*Psicopatologia del linguaggio. Storia, analisi, filosofie della mente* (Rome: Carrocci, 1998)), allows us to better understand the relationship between verbal and image-based language:

> Mirror-language is like a city: in it, public and private, exterior and interior are indissolubly linked but also autonomous. *Indissolubly linked*: every linguistic animal can have a private life only by observing itself in public discourses, both its own and those of others; the grammar of the common mirror is powered by the mirroring of each and every linguistic animal. *Autonomous*: the common mirror has potentialities that go far beyond the mirroring of individual linguistic animals; the specific mirroring every linguistic animal produces is not deterministically caused by the grammar of the common mirror. (15)

Since language is independent of any specific aim, the theory of the mirror-language allows us to contemplate in it the multifarious and complex multiplicity of modes, lifestyles and ways of thinking inherent in the human *Lebensform* (such as paranoia, schizophrenia, autism, and, we may add, dyslexia). The speed of dyslexic thought can be rendered with the example of the athlete:

> The athlete who runs five hundred meters in a few seconds is a walking two-legged animal, just like the author of these words. The locomotive apparatus of both bodies is governed by the same rules. The grammar of the locomotive apparatus prescribes some possibilities and impossibilities, that is to say, it traces a border within which both my slow walking and the athlete's fast running can be inscribed. (Ibid.)

9 Davis writes: "When we use the picturing process in nonverbal thinking we are not able to prefigure the meaning of the word as an object or an action" (R. Davis, *The Gift of Dyslexia* (Burlingame, CA: Ability Workshop, 1994) 13).

10 Here are some examples of trigger words in Italian: *a* [in/to] as preposition (*al allo alla alle ai agli*) [in the/to the] and as adverbial locution (*a stento*) [hardly/barely]; the word *anche* that can mean *also* as well as the plural of the noun *anca* [hip]; the verb *avere* [to have]; *che* [that], *ché* [because]; *quale* [which] as both adjective and pronoun; *questo*, *questa*, *questi*, *queste* [this, these], as adjectives and pronouns; the possessive *suo*, *sua* [his, her]; the verbs *vedere* [to see] and *venire* [to come].

11 M. De Carolis, *La vita nell'epoca della sua riproducibilità tecnica* (Turin: Bollati Boringhieri, 2004) 63.

12 Davis, *The Gift of Dyslexia* 124.

13 See Roger Succhielli and Arlette Bourcier, *La Dyslexie maladie du siècle* (Paris: Sociales Françaises, 1968). The authors define syncretic-analogical thought as follows:

> Syncretism is a form of thought and judgment that cannot be detached from the senses and that cannot think about relations among things. It is, clearly, an obstacle for the decisive objectivization of the universe and for abstraction. Only the latter can overcome analogies and constitute thought according

> to the analytical and logic model, as well as free the intelligence from affects and sensorial qualities. (56, Italian edition)

14 This detachment, this *separation* really, "allows a distance from the object, one that can generate the objectivity of the object, and makes the perception of the relations among the different parts possible, not just the perception of the relation between the pre-objective object and the pre-reflexive subject" (ibid. 63).

15 The idea of a mobile point of view may seem metaphysical, as if it were a sort of extra-sensorial perception. The theoretical attempts to explain this phenomenon range from, as already stated, the theory of *ectopia* to quantum physics, according to which perception itself produces some effects on the perceived object. According to other interpretations, there may be a form of perception that is not yet identified, just like the *sonar* that gives dolphins a tri-dimensional image of the environment and allows them to communicate these images to others. This could also be thought of as a form of imagination, where the individual mentally constructs multiple images of the object or symbol perceived. "People," Davis states, "naturally place their mind's eye in various comfortable positions. The mind's eye for dancers and athletes (two of dyslexics' preferred careers) is usually placed above their body, a convenient and advantageous position."

16 For orientation, the optimal position of the mind's eye is found through trial and error. It varies according to each individual and can change slightly with time, still remaining within a certain area. The localization is situated within ten and thirty centimeters above and behind the head, in the body's median line.

17 Davis, *The Gift of Dyslexia* 126.

18 According to some English researchers, the sense of rhythm is fundamental to learning how to read and write: dyslexia could be due to difficulty in perceiving the rhythm of sounds. Furthermore, an American study has suggested that the brain of dyslexic children could be "reorganized" with special computer training. To this end, software called Fast for Word Language was created in order to teach how to distinguish sounds that constitute words – children have to listen to them at different speeds. The stressing of the relation between auditory stimulation and spatial references allows dyslexics to activate a new elaboration through visual stimulation (see Raffaella Machinè, ed., "Conoscere la dislessia. Terapie e cure," *Azione* (3 Dec. 2003)).

19 Mark Crispin Miller's attempt to find something dyslexic in the idiocy that pervades the speeches of George W. Bush is utterly inconsistent and also offensive to dyslexics. This demonstrates, among other things, how dyslexia is still an unknown phenomenon (see M.C. Miller, *The Bush Dyslexicon: Observations on a National Disorder* (New York: Norton, 2001)).

20 For a description of some characteristics of digital capitalism, with concrete examples of emerging firms such as Charles Schwab, Ebay, Cisco, MP3, Linux and others, see the book by Don Tapscott, David Ticoll, and Alex Lowy, *Digital Capital: Harnessing the Power of Business Webs* (Cambridge, MA: Harvard Business School P, 2000). Particularly useful critical analyses of the Internet economy can be found in Carlo Formenti's *Mercanti di Futuro. Utopia e crisi della Net Economy* (Turin: Einaudi, 2002) and Franco Berardi (Bifo), *Il sapiente, il mercante, il guerriero. Dal rifiuto del lavoro all'emergere del cognitariato* (Rome: DeriveApprodi, 2004). For a critical evaluation of the new economy, see Rober Boyer, *La Croissance, début de siècle. De l'octet au gène* (Paris: Albin Michel, 2002); *Une théorie du capitalisme est-elle possible?* (Paris: Odile Jacob, 2004).

21 See Pierre Lévy, *Becoming Virtual: Reality in the Digital Age* (New York: Basic, 1998). "The virtual," Deleuze writes in *Difference and Repetition* (London: Continuum, 2004) "possesses reality as virtual." We can understand why the dyslexic Leonardo da Vinci could conceive of a submarine three hundred years before the invention of a mechanism that could pump water out of a vessel. His ability to think multi-dimensionally allowed him to experience his intuitive thought as reality and draw the results that everyone can see. Maurizio Lazzarato, in his book *Puissances de l'invention*, evidences Gabriel Tarde's critique of Darwinian positivism, based on the notion of the virtual:

> Tarde accuses Darwin of wanting to explain change in the light of the action of *actual* forces, that is to say, in the light of the "sole agents and forces that are known." The real we can know, however, cannot be explained unless it is "connected to the immensity of

possibilities." "The actual is only an infinitesimal part of the real," the "realized is only a fragment of the realizable." (*Puissances de l'invention* 65–67)

22 Davis, *The Gift of Dyslexia* 32.

23 John Chambers, Cisco's dyslexic CEO, when confronted with the incumbent crisis of the *new economy*, really tried to avoid firing twenty thousand of his precious employees. In the end, however, economic reason prevailed: employees were fired, he stayed.

24 It is important to underline that in new capitalism, in the "immaterial economy," what defines the market price/value is the exchange of *service*-goods, that is to say *access*. In the production of services, however, the *material* has a growing importance. To construct one laptop computer we need tons of hazardous material and fifteen hundred liters of water. The Internet is based on solid electrical industrial structures and the diffusion of servers has already changed the energy requirements of office buildings. And this is not to mention logistics, transportation of goods, oil consumption and its rise on a global scale, and of course the daily accumulation of waste, that is, all the *negative externalities* which allow us to lower the costs of production only by *subtracting value from the environment and the quality of life*. According to Alain Gras, we are effectively still in an industrial society, a "thermo-industrial" society, which is increasingly dependent on the combustion engine (see *Fragilité de la puissance. Se libérer de l'emprise technologique* (Paris: Fayard, 2003)). The immateriality of productive processes and of service-products is therefore referred to as the "putting to work" of human faculties such as linguistic-communicative and relational abilities, competences and knowledge acquired outside of the workplace (flexibility, affectivity, emotionality, reactivity, etc.), that is to say, the ensemble of human faculties that, confronted with automated and digitalized productive processes, are *directly* producing added value. In Marxian terms, we could add that the labor-power commodity is the *vital environment* as a whole, while *actual work* producing surplus value is the ensemble of the most general, that is to say *abstract*, human faculties.

25 This does not mean that private property will disappear. We are going through a process of concentration of riches that is increasingly extreme as well as through a parallel process of diffusion of stock properties. Rather, the juridical regulation of the property of social knowledge, in the form of intellectual property rights and patents, reflects the metamorphosis of the property form induced by the centrality of services and the access to immaterial services.

26 The contemporary technological system, based on new information and communication technologies, displays an atypical pricing structure, characterized by high fixed prices and, as a consequence, by irrelevant variable prices: price is practically independent of the quantity produced. What is costly is the conception of the product, while its production and distribution have a decreasing marginal price. Neo-capitalist companies benefit from economies of scale and *increasing revenues*: their unitary costs decrease while the economic results improve with the increase in their scale of production.

27 This is the *topicality* of Gabriel Tarde's critique of Schumpeter (who, nonetheless, has similar economic views) and of political economy in general. Maurizio Lazzarato highlights it in his *Puissance de l'invention*:

> The production of riches based on invention and cooperation has a remarkable property in contrast with political economy and Marxian criticism: it does not have a *center*. It does not develop, like in political economy and Marxism, according to a linear logic: production of value within the company, circulation of products in the market, destruction of riches and creation of value through consumption. The chain of production of value is parallel to fabrication, circulation, and consumption of the product. This takes place in all these places at the same time and also elsewhere, namely, in the field of science. Tarde includes the social, affective, cognitive and communicative dimension of inter-psychological relations in the cycle of value production and in the concept of wealth. This is the operation that political economy should solve, since "the theoretical and aesthetic sides of all commodities are increasingly developing above their unitary cost rather than at its expense." (44–45)

28 See the important revival of the Schumpeterian theory of innovation and its routinization in Baumol's *The Free-Market Innovation Machine*.

29 The computer revolution of services of the 1990s, Internet opportunities, and the speculation bubble abruptly changed the logic of innovation in the United States.

> It is a new version of the American dream, a new frontier where innovation is born out of the creation of bouquets of companies. Two persons, an idea, and a garage can create a new world firm, under the magic wand of risk capital. Microsoft, Amazon and Cisco fed this narrative. The beliefs of institutional investors from the US made them place increasing amounts of money in innovation funds. (Michel Aglietta, *Macroéconomie financière. I: Finance, croissance et cycle* (Paris: La Découverte, 2002) 33)

30 See Nathalie Moureau and Dorothée Rivaud-Danset, *L'Incertitude dans les théories économiques* (Paris: La Découverte, 2004). Michel Aglietta and André Orléan are among those who most rigorously developed the concept of mimicry starting from the information deficit that characterizes "ontologically" the subjects who act in the financial markets (see M. Aglietta and A. Orléan, *La Monnaie entre violence et confiance* (Paris: Odile Jacob, 2002); A. Orléan, *Le Pouvoir de la finance* (Paris: Odile Jacob, 1999)).

31 On the Internet convention, see Luca De Biase, *Edeologia. Critica del fondamentalismo digitale* (Rome and Bari: Laterza, 2003).

32 Thomas C. Schelling, *The Strategy of Conflict* (Oxford: Oxford UP, 1960) 97–98.

33 An often-cited example of a "bad" convention is the QWERTY keyboard on typewriters (where QWERTY indicates the order of keys). Paul David's work demonstrated that the choice of this particular keyboard is not the most efficient: another order of letters would have been better. Once elected, the QWERTY convention ended up prevailing over other, better conventions on the market. Keeping in mind that the QWERTY convention was chosen to *slow down* the speed of typists, in order to avoid frequent key jams, the *positive* effect of this "bad" convention was to allow everyone to use the typewriter – and now the computer – at their own speed.

HUNGER, REPLETION, AND ANXIETY

massimo recalcati

translated by wissia fiorucci

I the enjoyment of the monad

From a psychoanalytical perspective, one of the fundamental traits of the contemporary age consists in a generalized loosening of our bond with the Other. The current subject appears to be detached from the Other, adrift, deprived of those symbolic and ideal references that are meant to exercise a function of guidance; influenced by an unlimited offer of objects of enjoyment, inhuman partners which are at both hand and mouth reach, always available, and which have replaced the unpredictable contingency that characterizes our encounter with the Other sex. As many have noticed, the symbolic power of the big Other has irreversibly weakened, and our time is a time of monadic enjoyment, as Adorno already observed in *Minima Moralia*. In other words, it is the time of an autistic radicalization of the individual, which excludes the trans-individual dimension of the subject. Following Lacan, we must distinguish between the individual and the subject. First and foremost, because the subject is structurally split insofar as it implies an intrinsic presence of the Other (that is, of desire and of the alterity that structurally constitute him). As a consequence, the subject cannot be reduced to the individual who, as the etymology of the term tells us, stands for an undivided "being," an identity that corresponds to itself.

The enjoyment of the monad is, in this sense, a contemporary alternative to the subjective experience of desire as an opening towards the Other's alterity. In our time, the hegemony of the discourse of the capitalist tends to impose a new illusion, which is different from those promoted by religion and positivistic thought: *the illusion that the object of desire is embodied in the object of enjoyment*, that is, the illusion that, by consuming the object of enjoyment, it is possible to heal the wound that runs through human reality and makes it structurally precarious and lacking, hence subjected to the enigmatic power of desire.

The astonishing production of a whirlwind of objects and gadgets surrounds the hypermodern subject with an environment of collective mania. The bond with the Other and the dimension of the erotic-amorous exchange that it entails is substituted by the unilateral relationship with an unlimited series of inhuman partners (drugs, food, alcohol, psychotropic drugs, the fetishized imaged of one's body, virtual realities, and so on).

2 an impossible repletion

From this very general perspective, I consider anorexia and bulimia as two exemplary declinations of this substitution and of this new (post-human; monadic) version of the social bond. For the anorexic, her own idealized image becomes her fundamental partner; the world is reduced to the smooth and aseptic surface of the mirror. Hers is a passion of consistency: making herself identical to her ideal image, coinciding with that which, structurally, it is impossible to coincide with, that is, one's own ideal-ego, the narcissistic reflection reproduced by the specular image. Her mission is one of mastery: governing the body, exercising on it a domination of the will, a control of its chaotic appetites. What is really at stake in this task is a reversal that metapsychologically characterizes the obscene and ruthless figure Freud called the Super-ego (*Über-ich*). In the moral radicalization of mastering the will over the thrust of drives, the anorexic actually ends up making of this ascetic exercise the place of an unprecedented and paradoxical enjoyment. This is what, in *La Faim et le corps*, their major work dedicated to these issues, the Kestembergs appropriately defined as the "vertigo of domination" that unsettles the instinctual arrangement of the body as regulated by symbolic castration, and elevates the renunciation of the drives to a full-blown lifestyle, an inverted form of dandyism that works as an aestheticizing regulation of the drives. Nonetheless, clinical experience shows that the anorexic prolongation of abstinence and of its multiple strategies of control inevitably tends to generate uncontrollable bodily phenomena. This is, for instance, the case with the production of endorphins that overwhelm the subject with a flow of excitation which is as powerful as that conveyed by raging hunger. This excitement is produced by means of exhaustion, the fading out of the instinct, and further clarifies how in anorexic fasting the body is not just obliterated, but rather enjoys the object's absence as if it were the object at its fullest. This unnatural enjoyment forces us drastically to revise the very idea of drive satisfaction as dependent on the fulfillment of an urge, an inner tension, such as that which is caused, for instance, by the emergence of hunger as a need. Rather, there seems to exist an *enjoyment of the void* that is stronger, or, more precisely – to use a Freudian expression – "closer to the drive" than that of repletion.

In order to understand this argument further, we should address the other side of anorexia, that is to say, the bulimic experience of hunger. Differently from anorexia, what comes to the fore in the case of bulimia is the pure and positive activity of the drive, and not its fanatic interdiction or its nirvanic sterilization. However, although the activity of devouring, as if one were starving, replaces the ascetic refusal of the anorexic-restrictive type, just like in anorexia, what matters the most even in this case is not the object that is being eaten but the very activity of eating. It is as if this very activity excluded the concept of repletion since it turned out to be totally independent from the dimension of the natural satisfaction of a need. To put it differently, we are not dealing with a hunger directed at the object, but at the void as the impossible core of the object. For this reason, bulimic hunger does not literally know the sense of repletion; it does not want to feed the subject with food but with enjoyment, the enjoyment of that which cannot be grasped, the lost Thing, the fantasy of the breast.

On the other hand, this compulsion to eat everything without ever attaining an authentic repletion amounts to the point at which bulimia converges the most with the new, social dictate that regulates the contemporary program of civilization: what really matters is not the consumed object but the very activity of consumption, the "consumption of consumption," following Baudrillard's formulation. This highlights the purely drive-based content that inspires the discourse of the postmodern capitalist. This is, so to speak, the other side of the contemporary Super-ego. If the fundamentalist hygienism of anorexia leads towards a radicalization of the drive that animates the Kantian will, the bulimic devouring rather exalts the sadistic commandment to enjoy as a new form of the social Super-ego.

3 the empty subject

In *Clinica del vuoto: anoressie, dipendenze e psicosi*,[1] I referred to a *clinica del vuoto* [*clinic of the void*] to define an anthropological metamorphosis that has invested the so-called postmodern subject. This metamorphosis tends to reduce the subjective dimension of the "want-to-be" (*manque à être*) and that of the desire that originates from it – isolated by Lacan as the two polarities constituting the very being of subjectivity – to that of a void detached from lack and deprived of any bond with desire. Therefore, we must distinguish the void of the "clinic of the void" from the lack as the effect of the cut operated by the signifier on the body of the subject. The contemporary reduction of lack to the void entails, first and foremost, an effect of false mastery. That is to say, lack turned into the void offers the illusion that the latter can be filled, as happens in bulimia or in other forms of pathological addictions (for instance, in obesity or in drug addiction), or the illusion that, as happens in anorexia, it is possible to "ossify" the void, turn it into something that is full [*rendere pieno*], and make of it the subject's permanent center of gravity – and of enjoyment.

From a phenomenological perspective, anorexia and bulimia represent the two sides of the same coin. We write "anorexia-bulimia" with a hyphen to indicate their mutual belonging. Anorexia is, from this point of view, a virtual bulimia, while bulimia is the destitution of the anorexic's endeavor to dominate the drives. If we could further explore this argument, we would perhaps find in this cyclic alternation of anorexia and bulimia a point of intersection, which the Lacanian figure of the object *a* would help us to circumscribe. All things considered, in order to get as quickly as possible to the heart of the matter [*l'osso della questione*], we must acknowledge in both oscillations the same void point that characterizes the activity of the oral drive, and that, in Lacanian algebra, is one of the decisive meanings ascribed to the object *a*.

What happens then to the void which is at the center of the human activity of the drive? The anorexic incorporates it; she must feel it in her stomach; she actively seeks it by refusing to eat, or, better still, as Lacan claims, by eating the "nothing" (*rien*). The bulimic, on the other hand, can only find it at the apex of her enjoyment: she devours everything and stuffs herself with it, but only to be able to show the illusory character of repletion and, thus, to try and reach the void of the breast-object (which is impossible to reach). She would like to take a bite at the void of the Thing, the fantasy of the breast, that void which is impossible to eat, around which, however, the oral drive always revolves.

More specifically, the anorexic seems to establish an original tension between alienation and separation. My thesis is that there is a *separation-against-alienation* in anorexia. What does it mean? The anorexic invokes and practices, in an apparently radical way, separation. First of all, this is the separation from the demand of the Other and, more generally, from any possible form of demand. In fact, she does not demand anything and refuses everything. On the other hand, this separation seems to be produced without a loss – which is the structural effect of the signifier's basic grip over the subject; the object *a* seems to remain on the side of the subject rather than being transferred to the field of the Other. This is the extremely determined trait that characterizes the anorexic choice. The necessity of separation originates from the negation of loss and not from its subjective assumption. In a possibly more Kleinian vein, we could claim that, in anorexia, separation takes place without being accompanied by an effective work of mourning. It is a separation that takes place by means of the will, not desire.

4 anorexia, bulimia, and anxiety

Generally speaking, one could say that anorexia is basically deprived of anxiety, because it takes the shape of one of its treatments, of a particular strategy to avoid anxiety.

If, for Freud, Lacan, and Heidegger, anxiety is that affect which makes the encounter with the real possible (the real of the drive or of existence), the anorexic would like to demonstrate that it is possible to live our existence without anxiety. In doing so, anorexic apathy aims at subverting the structural character of anxiety. If, from a

Lacanian standpoint, anxiety signals the emergence of the object *a* as an indication of the drive-based character of the body, anorexia would be an attempt at cementing the image, in order to subtract it from the perturbations of anxiety and reassert by this means an idea of the subject as pure identity. Being is solidified by rejecting the signifying alienation. If, for Lacan, anxiety manifests the real depth of the body's drives, anorexia erects an imaginary dam that intends to hide this inner otherness. For the anorexic, the only possible experience of otherness is that of food and calories. The drive re-emerges only in its *bulimic decompensation*, shattering the dam of the ideal-ego and exhibiting the acephalous character of the drive's movement. For this reason, the subjective terror that animates the anorexic is that of experiencing the bulimic aspect of her hunger. In this manner, the sterilized and nirvanized body of the anorexic traumatically re-acquires the drives of the body which, like an inner "Alien," can never be suppressed; after all, they coincide with the very life of the subject.

From this perspective, the anorexic tries to challenge Heidegger in a Freudian way: *it is not nothingness but the hunger of the drives of the body that causes anxiety*. For this reason, the anorexic's endeavor consists in the realization of an egoic-voluntaristic mastery of the body, as opposed to the panic that, not accidentally, can arise only when the anorexic ascesis diminishes. It is a matter of pushing away the object of anxiety. For this reason, we often find in our patients infantile phobias involving certain types of food, and we could even pose the more general, psychopathological problem of the relationship between anorexic practices and phobic-obsessive systems . . .

From a bulimic perspective, the experience of anxiety is one of libidinal clogging. The subject is suffocated by what she eats. Anxiety does not arise from the lack of an object, but from a real excess of it, precisely from an experience of oppression, lack of lack, stuffing of the body, asphyxia. Therefore, having recourse to vomiting often coincides, in the patients' life, with the recuperation of their body, with the possibility to create the void anew, to reopen that little gap which allows the re-evocation of desire. The force-fed-body, the body glutted with food, the full-body [*corpo pieno*] is, on the other hand, the body of anxiety. This is why in the dreams of many anorexics this body that "overflows" ["*troppo pieno*"] appears in all its monstrous significance as a materialization of the chaos of the drives, which is impossible to control, of the body as an alien, as an animal or extraterrestrial body.

The task of mastering one's body is carried out by the anorexic as an idealization of the specular image that would like to annihilate the potential apparition of this alien monster. Drawing on Lacan, we know that one of the fundamental functions of the image is that of covering the body of the drives by offering it some identifiable borders (for Lacan, the schizophrenic is somebody who does not have access to the imaginary). When, conversely, the body loses its image, we can observe the appearance of different clinical phenomena, for instance, that of melancholia, which is of great importance in the clinical approach to anorexia-bulimia. Without a narcissistic image, the body emerges as a pure object of waste, as a *kakon* object, as an ugly and real body deprived of the very feeling of life, as a body that is already dead.

When the anorexic control loses its pace, anxiety can find its way through. This infiltration signals the narcissistic image's impossibility of entirely covering the body of the drives, as a consequence of which a separation between the narcissistic and the erogenous body takes place. For this reason, in looking at oneself in the mirror, what produces anxiety is always dependent on an excess of flesh, or fat, which smears the beautiful image. It is some sort of residue that embodies the object *a* as something that cannot be specularized, as a real that we cannot integrate in the good form of the imaginary. It is for this reason that the age of puberty continues to be the best time for the triggering of anorexia. During puberty, it is the real of the body which comes to the fore tearing apart the unity of the beautiful image. The anorexic refusal is, first and foremost, not a refusal of the oral object, but a refusal of the body *tout court*. It is a refusal that usually takes the path of a refusal of one's own body as sexual body and a refusal of the Other's body as a place of enjoyment and desire.

5 the two souls of anorexia

In *L'ultima cena*,[2] I advanced the idea of anorexia as a love-sickness. To get hold of the sign of love, of the Other's lack, the anorexic subject chooses a desperate path, that of a radical refusal of enjoyment. Here, we could isolate the hysterical trait of anorexia. It seemed to me that bulimia itself was oriented by the same sickness: as Lacan recalls in *Seminar IV*, the absence of the sign of love is made up for by devouring the object. In *Clinica del vuoto*, I deemed it necessary to highlight another dimension, that of hatred, of the refusal of life not as a call for love but as a death thrust [*spinta alla morte*]. There is a great difference between anorexia as a call for love and anorexia as an appetite for death [*appetito di morte*], one that is not sufficiently clarified by Lacan in his notes on anorexia, which nonetheless always appear at topical moments of his teaching. For instance, anorexia is evoked as a key clinical figure that gives access to the category of enjoyment, as is already the case in *The Family Complexes*, but also to the category of desire, which is the case in *Seminar IV* and in the article "The Direction of the Treatment." Synthetically, Lacan offers a sort of double reading of the anorexic phenomenon: on the one hand, he describes it as the place of a deadly, melancholic, and addictive enjoyment, on the other, as a strategy of defense and separation of the subject's desire from the suffocating demand of the Other. He emphasizes the double soul that characterizes the anorexic subject as such: manifestation of the *Todestrieb*, appetite for death, larval desire, destructive thrust, melancholic annihilation of the feeling of life, nirvanization of the pleasure principle, but also strategy of separation aimed at differentiating the status of desire from that of need, irreducibility of desire to the demand of the Other, desire as desire for nothing, desire of the Other, refusal as a radical form of call, love-sickness, radical demand of the sign of love.

If love makes desire and enjoyment converge, anorexia and bulimia oppose desire and enjoyment and exclude the conversion of love. Anorexic desire is in fact a desire for death, and bulimic enjoyment can appear not only as a form of compensation but also as a devastation of the drive.

notes

1 M. Recalcati, *Clinica del vuoto: anoressie, dipendenze e psicosi* (Milan: Angeli, 2002).

2 M. Recalcati, *L'ultima cena* (Milan: Mondadori, 1997).

pietro bianchi

edited by guillaume collett

THE WORD AND THE FLESH

postworkerism and the biopolitics of language in paolo virno and christian marazzi

I

When one views the whole panorama of the Italian postworkerist tradition, Paolo Virno in many ways stands out as a double exception. On the one hand, throughout his whole life he remained one of the closest and most faithful partisans of this theoretical and political tradition (an example of this is his stubborn attachment to thinking the category of the multitude even in his most recent works, despite the relative decline in usage witnessed by the term after its heyday in the early 2000s). On the other hand, within this tradition he is probably the member who is the most unscrupulous about inserting non-orthodox references, to the point that in his latest books his references include Wittgenstein, Vygotsky, Gehlen, Plessner, Benveniste, and even Vittorio Gallese and contemporary cognitivism, all revisited in a coherent and highly individual fashion. Paolo Virno is at the same time an undoubtedly symptomatic representative of Italian postworkerism (with all the defining elements pertaining to it), and a highly original thinker who does not resemble anyone else from his generation or from the tradition of thought to which he belongs. In terms of originality, something similar could be said of Christian Marazzi: among Italian postworkerist thinkers, he has been *par excellence* the one who has devoted all his efforts to a serious and even sometimes technical analysis of the different economic and financial cycles of the last two decades. Marazzi's seminal and deeply innovative book *Il posto dei calzini* [*The Place of Socks*] published in Switzerland in 1994 (he is from the Italian-Swiss region of Ticino) probably constitutes *the* inaugural breakthrough moment for the analysis of the contemporary form of post-Fordist labor and the current process of production. Marazzi is the only true economist among the postworkerists, a tradition whose theoretical efforts, at least in the last twenty years, have almost always been confined to disciplines such as philosophy or sociology. This is not an insignificant consideration for a theoretical and political position that still openly refers to Karl Marx. It is also important to note that in the last few years, and especially since the dot-com bubble, Christian Marazzi's work has been increasingly focused on the study of financial processes and especially on the role that linguistic practices play in them. Language is already considered to be a crucial economic factor in any sociology of labor study that analyzes the changes in the organization of production since the so-called post-Fordist turn in the 1980s.

But the way language also deeply affects the realm of finance has somehow remained under-theorized. Marazzi has developed an analytical framework that decisively underlines the increasingly crucial role played by financial capital in contemporary forms of accumulation, but, even more importantly, as a consequence of that, he has been able to stress how the contemporary extraction of value based on financialization relies on the valorization of the common (and mainly linguistic and relational) features that define all speaking beings.

In fact, one could be surprised to see such interest in and refined speculation about the invariant and species-specific nature of language, and its relation to human nature, coming from a resolutely Marxian point of view: we should nevertheless note that Paolo Virno's and Christian Marazzi's interest in these topics was not an isolated development in postworkerist thought; at the same time, it certainly does not stem from the 2000s when, for instance, some of these issues became relevant to the highly popularized thought of Antonio Negri, regarding the multitude or Commons. A certain focus on the specific nature of language and its relation to contemporary forms of capitalist accumulation was already present long before, at the beginning of the 1980s. And it was deeply connected to the transformations in the capitalist means of production that occurred between the so-called Fordist stage of capitalism, centered on the figure of the mass-worker [*operaio-massa*], and the so-called post-Fordist stage, centered on the figure of the social-worker [*operaio-sociale*]. The postworkerist tradition of thought was deeply involved in a resolute (to the point of being daring) effort to test some of these transformations using a highly innovative Marxian (or to some extent even post-Marxian) perspective.

Around the mid-1980s, this new generation of intellectuals and militants, more or less related to the many waves of the Italian Workerism of the 1960s and 1970s, started a new phase of analyses focused on the transformation of labor and of the production process. Paolo Virno and Christian Marazzi, along with many others, played a very important part in this highly innovative theoretical effort. Firstly, with the journal *Luogo Comune* and then from 1991 with the journal *DeriveApprodi* (which lasted until the counter-globalization movements of 2001). The idea (pursued with tenacious recklessness) was that the development of labor *tended* towards a situation where the most productive part of labor was increasingly constituted by a "general intellect" and by highly developed forms of social cooperation, leaving material industrial labor with an increasingly marginal role. This thesis had a somewhat theoretical basis in the famous (or infamous) and rather historicist Marxian passage from the *Grundrisse* entitled "Fragment of the Machine" (a passage that became a sort of "myth" in the workerist literature), which is where the term "general intellect" (in English in the original text) appears for the first time. In this text Marx in fact seems to predict a possible stage in the development of capitalism where, through the form of its incorporation into fixed Capital, knowledge (and especially technical and scientific knowledge) constitutes the primary productive force.

Even though the macro-economic accuracy of the "general intellect" thesis could be easily relativized (and it actually has been, in a very harsh debate within Marxist Studies that is still going on), it nevertheless had the great merit of underlining certain descriptive features of contemporary labor that questioned the Marxist doxa that still relied on a stereotypical representation of labor centered on the Imaginary of the Fordist and industrial working class. In this regard, the postworkerist tradition undoubtedly constituted a vivid and important intellectual effort to widen the boundaries of what is and what is not considered labor. The political analysis of the environment of small entrepreneur and autonomous/freelance workers (which were traditionally very de-politicized and far from the values of the left) served to update the analysis of a production process that at least in some regions – such as the northern metropolitan areas of Italy – no longer had the material production of commodities at its core, but instead a general supply of cultural and financial services. The predominant linguistic features of this highly developed and tertiary-oriented form of labor also led to the coining of the heavily contested phrase "*immaterial labor.*"

This premise is important in helping us to understand which kind of question the analysis of the invariant features of human language is trying to answer. Virno, for example, is not trying to frame a purely philosophical problem: in fact, first and foremost, he is interested in analyzing the way in which a contemporary form of capitalist accumulation is developing itself. But in order to do that it is necessary to take a *detour.* According to him, it is in fact especially in *these* historical times, and particularly with the transition to a post-Fordist mode of production, that some generic faculties of the human animal start gaining value and becoming prominent: knowledge, relationality, adaptability, non-finalized capacities, and so on. Therefore, it is because of political urgency (because it is crucial to have the most refined analytical tools to understand *this* mode of production) that we need to focus our attention on the invariant and generic features of the human animal. We need to develop an anthropologically oriented *political* analysis. Even when Virno is talking about Wittgenstein or Aristotle, he never ceases to be a political thinker. It is not that anthropology *per se* plays an important part in helping us to understand *any* mode of production. It is – with a formula which became almost a slogan – *only now* (in the post-Fordist mode of accumulation) that what was *already there* (the biologically invariant and transhistorical features of the human animal) has gained uncommon importance. To put it simply, it is only in the contemporary stage of capitalism that class struggle revolves around the question "What is human?," because (and we will see this in the work of Christian Marazzi) value is extracted mainly from human nature itself (the linguistic and generic features of every speaking being) in this stage of the accumulation of capital.

2

According to Paolo Virno, the human animal, in a very paradoxical formulation taken from Gehlen, is defined only by its lack of definition, by its lack of specificity. It is an undetermined animal. The human animal is not able to perfectly orientate itself in its environment, while all other animals occupy a certain place in their environment. The human animal is literally out of place. This generic *a priori* condition appeared alongside the post-Fordist mode of production and so it is also a historical condition. What before was only a biological pre-condition (a potentiality expressed differently in every historical phase) became, with post-Fordism, a directly (or *immediately* as the postworkerists like to say) historical one. We can even go so far as to say that, according to Virno, the contemporary form of accumulation collapses the boundaries between biology and history (an end of history?). This lack of specificity should not be considered as the deprivation of an idyllic past, but rather as the final appearance of the true generic substance of all human animals. Virno claims that when we hear the ideological post-Fordist rhetoric of being ready for constant innovation, relationality, linguistic abilities, etc., we are provided with nothing other than the perfect definition of the human being as an undefined animal at the center of a historical process. In this sense, according to Virno, post-Fordism is – surprisingly enough – *the* historical time when the affirmation of the genericity of the human being proper to communism is finally becoming possible.

In *Multitude: Between Innovation and Negation,*[1] we find three different essays by Paolo Virno. The first one originally appeared in 2005 in volume 4 of the journal *Forme di Vita* [*Forms of Life*] dedicated to "The Dangerous Animal: Human Nature and Political Institutions," and it is a discussion of the way in which political institutions deal with the aggressiveness and instability that characterize the speaking species. The second essay is the whole of a short book published in 2005 with the title *Motto di spirito e azione innovativa. Per una logica del cambiamento* [*Jokes and Innovative Action: For a Logic of Change*], which, starting with the Freudian analysis of the *Witz*, analyzes the relationship between linguistic rules and practices which suspend them, leading to the production of the new and unforeseeable. The third essay, "The Innovative Action. When a Form of Life Changes," was published in 2004 in volume 2–3 of *Forme di Vita*, and focuses on the primary intersubjectivity implied by the scientific discovery of mirror neurons.

We will start our analysis with the second part, which, even though in this volume it is put after the discussion about human language and the institutional problem, is in fact its premise and should be treated first. This chapter ("Jokes and Innovative Action: For a Logic of Change") revolves around the question of the production of the new and the way in which the practice of novelty in the human animal can rely on logico-linguistic resources. Unlike Chomsky, who proposes a natural language which is constantly and intrinsically innovative, Virno struggles with language's inevitable intermittence: why, in which cases, and under what circumstances, can innovative linguistic action take place? Why is it that in some cases the human animal only has recourse to the repetitive and stereotypical set of rules of language, while in other cases it is able to expose language's paradoxical short-circuit, its internal contradiction, which itself produces an unforeseeable and innovative action? This essay uses the example of Freud's famous analysis of the procedure of the *Witz* in his *Jokes and their Relation to the Unconscious*,[2] published in 1905. Virno claims that he is not interested in any psychoanalytic interpretation of it, but we will see that in the end he gets very close to many problems and issues that are prominent in the psychoanalytical debate, something which will unfortunately remain in the shadows. The two main interesting aspects of the *Witz*, according to the Italian philosopher, are its creativity (Freud – it is recalled in the book – said that jokes are "made," "produced," as opposed to simply comic situations, which only need to be recognized as such in order to be comical), and the public space that is needed: in order for jokes to take place it is not enough to have an author, it is also necessary to have a "third" person who constitutes the audience of this very joke, and who laughs at it. Without an audience it is impossible to verify whether the joke has hit its target. Virno is underlining another crucial Freudian point:

> The "tendentious" remark, consisting of aggressive or obscene content, and, equally, the "innocent" remark, that restores the childish habit of playing with words as though they were things, require of the person who coins the remark a conspicuous waste of psychic energy in order to overcome all sorts of inhibition [...] For makers of jokes, the work entailed in making something new (and not agreed upon) erodes and neutralizes their eventual "profit or pleasure."[3]

In fact, even though the third person shares the same inhibitions as the joke teller, the former is nevertheless the one who is able to enjoy and laugh without any psychic waste. This latter element constitutes an important point of departure, enabling Virno to underline how the practice of joking does not constitute an inner transgression (a private matter that pertains only to the egotistic enjoyment of the joke teller). It is rather a *public* action, nothing less public – as he says – "than a political discourse held in a general assembly that urges towards the insurrection against the constituent powers: if enunciated in the absence of witnesses, it is as though this discourse had never occurred."[4] It is a linguistic action that must not be measured by the accuracy of the content uttered, but by the changes that can be produced in a performative manner, even though it differs from more usual performative utterances precisely because it lacks the ritual dimension that is inevitably present in juridical rituals, public institutional discourses, and so on. Therefore, we must consider the performative and collective aspects of the joke (the role of the third person as an audience) to be inseparable from the innovation made possible by the joke teller.

But how can a joke constitute a transformative act capable of rearticulating a set of rules, and not just bring about an inner transgression that will end up reinforcing the rule? According to Virno, jokes are linked closely to one of the most insidious problems of linguistic praxis: the question of the relation between rules and particular cases. Far from being the pure application of a general rule to a single case, a joke allows us to judge situations that completely lack the frame of reference provided by the legislation of the Law and which therefore require a special regulation (a state of exception). It is what Aristotle defines in the *Nichomachean Ethics* as *phrónesis*: the ability to act in situations which have a high degree of indetermination, and where the right thing to do is not implied in the Law. Virno is underlining how, in Aristotle himself, there is a certain circularity between the logic of exception (the insightfulness of *phrónesis*) and the correct discourse of the Law (*orthós logós*): a discourse establishes itself as correct (*orthós logós*) only if there is the practical shrewdness of *phrónesis*, but *phrónesis* itself exists only on the condition that the proper norm is enunciated. *Phrónesis* therefore exposes the logic of the punctuating enunciation implied in every linguistic act. As everyone knows, in order for a joke to be effective it is necessary to pick the right moment, to seize the *kairós*: one moment too soon or too late would cause the joke to inevitably fail. In regard to this, Freud mentions the term "involuntary idea": something that seems to abruptly come out of the pure shrewdness required to deal with the situation. This something has a very different structure from a judgment or an argument, which still seem to rely on the dual temporality of the pure existence of a potentiality (the enunciated content) and its moment of actualization (the moment when the enunciated content is uttered in the event of the enunciation). In relation to this, Virno quotes this very important passage from Freud: "What happens is not that we know a moment beforehand what joke we are going to make, and that all it then needs is to be clothed in words."[5] To which Virno adds:

> Precisely. But this observation holds not only for the joke: strictly speaking, it *never* occurs that there is an already constituted thought that afterwards will be embellished by words. The joke, precisely because it is motivated by the necessity of seizing the moment, demonstrates with exemplary precision what is always true: the integrally *verbal* character of human thought. The fact that we think with words becomes empirically evident (I will know what I was thinking only after having said it!) when the process of thinking-speaking executes an innovative action, an action that, in order to accomplish its goal, must take place *precisely now*.[6]

The Law, the rule, the *orthós logós*, thought as a pure potentiality, cannot exist outside of the precipitation of an act: the punctuating moment of enunciation. There is no *epistéme* without the moment of *praxis*.

Here we start to glimpse what is at stake in this analysis of the Freudian *Witz*. Virno is not so naïve as to think that a joke could really involve a true transformative and political act; rather, he wants to tell us that it could serve as a *diagram* which can be used to enlighten a more general principle:

> Jokes are the logicolinguistic diagram of the undertakings that, on the occasion of a historical or biographical crisis, interrupt the circular flux of experience. They are the microcosm in which we witness clearly those mutations of deductive direction and those displacements of meaning that, within the macrocosm of human praxis, provoke the variation of a form of life.[7]

Jokes can expose the strict relation that entangles *epistéme* and *praxis*, rule and application, Law and state of exception. But instead of being haunted by a foundational problem (as is the case in some of Agamben's works), Virno seems to be more concerned with a *pragmatics* of change: within a certain given set of rules, how can a form of life expose its contingent determination to itself, and therefore its own practice of rearticulation?

Between rule and application there is a "logical fracture," an incommensurability. At a certain point every rule exposes itself to the multiple alternatives it has available to it regarding how one can apply it. When we are confronted with a rule, there is never a meta-rule that will guarantee its mode of application. In this regard, the joke represents the way in which, when one is confronted with a rule, different and contradictory ways of applying it can be explicitly carried out, consequently exposing the fact that every application cannot help but be derived from a non-founded act of decision. When in front of a road sign, instead of continuing down the road we can decide to take a side road – or to go across the fields – we expose the impossibility of founding a norm on something other than a contingent act of decision. On the contrary, if we try to base the application of a norm on another norm, we will easily end up in a vicious circle where the primary norm requires another one for it to be founded on, then another one ... and so on to infinity. A decision in fact means that we cut this never-ending circle open (in Latin, *caedere*, which means "to cut," is the root of "decision").

So, in terms of applying a rule, if contradictory possibilities always co-exist, how do we decide which one to choose? Which is the right one? How should we act when the rule is suspended momentarily (and if we take Virno's argument to its logical conclusion, this is something that is microphysically present in every aspect of human behavior)? Here, in order to get out of this cul-de-sac, Virno ambiguously refers to a passage from Wittgenstein in which the Austrian philosopher evokes a hypothetical explorer, who, when faced with a language that he cannot understand, starts to behave according to what he refers to as *die gemeinsame menschliche Handlungsweise*, "the common behavior of mankind." What does this common behavior look like? Here Virno mentions the "distinctive traits of our species," "the fundamental dispositions of the linguistic animal":

> The natural history of which Wittgenstein speaks, almost at the beginning of the *Philosophical Investigations*,[8] in Section 25: "Commanding, questioning, storytelling, chatting [but also from §23, forming and testing a hypothesis, guessing riddles, making a joke; telling it, thanking, cursing, greeting, praying] are as much a part of our natural history as walking, eating, drinking, playing."[9]

When faced with a critical situation, with a linguistic "state of exception," the Wittgensteinian explorer – according to Virno – "come[s] back *to this side of* the rule." It is a "nonjuristic order [. . .] characterized by a radical blurring of fields and contexts," where "there is no other rule but the application." This other side of the norm is defined by Virno as *regularity*, the pure externality of every rule, and he is even unafraid of describing it as "a *pure application*, antecedent to any positive norm."[10] The problem with this kind of account is not just the ambiguous substantialization of a "common behavior of that mankind" antecedent to every rule but also the fact that this very line of argumentation is the direct consequence of a (previously addressed) problematic relation between rule and application. It is in fact very easy for the incommensurable space between these two to change from a *logical* space to a *temporal* one. Rule and application are not only incommensurable but also mutually implicated; furthermore, their mutual implication results from none other than their incommensurability (in a Hegelian way, only terms which are mutually implicated can be incommensurable). Here, Virno is repaying the debt owed to psychoanalysis, which he had dismissed at the beginning of the essay: it is in fact structural psychoanalysis, and in particular Jacques Lacan's theory of the signifier, that can help us to better frame the restlessness of the Law, and its always contingent point of inscription (the application).

The relation between Law and application is in fact structured around a central void: that is to say, the inexistence of a meta-rule that would have resolved the problem of its own foundation. Virno effectively underlines how the infinite regress resulting from this foundational question can go on for ever. It is only with the unexplainable moment of decision that the rule can become actual. This moment, far from exposing a certain lack in the rule itself, represents rather its central (even though external) core (what Lacan defined as *l'extimité*). This point is actually not even only Lacanian, being fully deducible from Saussure himself: *langue*, as a system based on purely negative differences, entails a self-erasure of itself without a self-reflective point of enunciation. A "pure," working structure would logically be a self-effacing one. It is because of this contingent point of inscription (what in Lacanese is defined as the master-signifier, or the external utterance of *parole*) that the system of language can actually come into being. It is here that we see the paradoxical overlap between incommensurability and mutual implication. In a system like language, every single element is defined by its *place*, which is none other than the negative difference subtending each and every other element (in structuralist terms an element is co-extensive with its own place). A purely differential element therefore does not logically have any positive substance whatsoever. How, then, can something that does not have any substance come into being? How can a structure – which if it were purely self-established would efface itself – come into existence? The answer is: by encountering a pure contingent point of inscription (a pure enunciation), like the *pure potentiality* of a voice (before any specific utterance). A pure enunciation is only logical and therefore it cannot be substantialized, given that it would be equally inexistent without a system of pure differences to which it could attach itself. The emergence of language is therefore *a paradoxical nothing that encounters another nothing and is consequently able to create something*. The incommensurability of law and application (or enunciated and enunciation) is nevertheless a figure of their mutual implication without which they would erase themselves. Contrary to this, Virno gives an account of the regularity that emerges in the blind spots (and therefore in the absence) of a rule (he frequently uses the term "suspension" of a rule), and consequently ends up addressing the notion of a biological invariant that risks relying too much on a substantialization, as is clear from the aforementioned passage taken from Wittgenstein. Here we have a conflict between two different notions of the *genericity* of human nature: on the one hand, we have a procedural genericity which is co-extensive with a purely negative act of separation that can never reinscribe itself in a substance; on the other hand, as in the case of Virno, we have a notion of human nature that, in the end, even though it is defined as a "pure enunciation," nevertheless appears to be based on historical evidence (neoteny, pure potentiality of speech, relationality, etc.). This is even more striking if we go back to the very premise of Virno's theory, that is, the "historical appearance," *especially now*, of the pure human potentiality that was *always already there*. The very substantialization of human nature, therefore, is none other than the untheorized starting point of, and taken-for-granted historical evidence for, the transition to post-Fordism. From being merely a change in the organization of production, this event becomes a quasi-mythical transformation which is even capable of turning pure potentiality into historical actuality. Ironically enough, from this perspective, post-Fordism ends up not being very far from the highly deprecated idea of the "end of history."

If we now jump from the second part of the book to the first, we see the translation of this anthropological problem into political terms centering on the question of the institution. Virno underlines how every theory of political institutions always relies, explicitly or implicitly, on a certain understanding of human nature (as in Thomas Hobbes where the political anthropology of *De Homine* constitutes the necessary prerequisite for the subsequent reflection found in the *Leviathan*). Virno takes a detour through Carl Schmitt in order to stress the connection between the allegedly Hobbesian "evil" character of human nature, and the subsequent need for the State to repress it through its monopoly on violence. At the same time, a political position of harsh criticism against the State, such as that developed by libertarians like Noam Chomsky, is based on the prejudicial idea of the "natural goodness" of our species, with the centralized apparatus of the State humiliating the inborn creativity of verbal language.

Beyond the alleged goodness or evilness of human nature, according to Virno, what we are tackling here is nothing less than the *radical ambivalence* of linguistic negativity. A human being is able both to act maliciously and to invent something radically new because of the unpredictability and uncertainty given to him by his lack of specification, which is language's *mark* on human biology. The overwhelming biological stimuli not connected to any specific operative task are able to produce a constant "opening": a radical detachment from any defined environment, and an inability to predict whether an action will be positive or negative. It is because of language that human animals are able to defend themselves against the radical evilness of animality, but at the same time this produces even more dangers. The human animal's ability to be "open toward the world" makes him able to defend himself against evil, but at the same time it makes him the agent of this very evil.

So, how do we address the question of political institutions without relying either on the naïve libertarian trust in innate human "goodness," or on the disenchanted Schmittian/Hobbesian notion of an "evilness" which will inevitably call for the centralization of political power? A good way to frame an answer can be found in Hobbes himself: in the problem of the articulation between the pre-juridical natural life and the "unified political body" that exercises a monopoly over political decisions. Virno here underlines a catch: *obedience to the Law is foundational*, i.e., it is required *before* any system of positive norms can be made explicit. We are asked to obey before we know what we will be ordered to do. There is no trace in any legislation of the meta-rule that commands our obedience to the Law. Therefore, there is no temporal relation (first the former, then the latter) between the "state of nature" and the "unified political body," but rather a mutual implication that makes Virno claim that any escape from the state of nature is logically impossible. The "state of nature" haunts the State itself from the inside, in the form of a repetition of the never-ending problem of its applicability. There is no meta-rule that would be able to inscribe the Law, whose application cannot help but be based on the pure contingent dimension of the act. Here Virno (as in the discussion of the relation between rule and regularity), instead of leaving this logical paradox open, cannot avoid the temptation of giving an account of those moments when the "state of nature" erupts, disrupting the consistency of the "civil state": in the form of (1) the *multitude*, or (2) the *state of exception.* In the last section of this chapter the (relative) contemporary decline of the sovereignty of the modern State is taken as an example of the fact that what was *always already there* (i.e., the self-reflective moment of the inscription of the Law in the form of the "state of nature") has *only now* become a possible way of experimenting with an institutional form proper to the multitude. In this conjuncture, the ambivalence of language as a species-specific anthropological *a priori* (beyond good and evil) and the pure enunciation of the Law will serve as the basis for a historico-natural political institution.

In the short third part that concludes the book, "Mirror Neurons, Linguistic Negation," Virno again develops the point regarding the ambivalence of language. This time the scientific discovery of the mirror neurons (in this case mediated by the work of Vittorio Gallese) provides the background to his discussion: these neurons are activated in the same way, whether an animal is acting or whether it is observing the same action performed by another (which is taken as an example of the fact that the potentiality of recognition is directly inscribed in the biological apparatus). Language, according to Virno, actually entails the opposite: it *negates* this primordial recognition. Language makes the failure of mutual recognition possible, tearing apart this pre-linguistic form of empathy:

> Every naturalist thinker has to account for the fact that the human animal is capable of not recognizing another human animal as a fellow creature. This permanent possibility is evident in extreme cases, such as cannibalism and Auschwitz, but it mainly manifests itself in paler shades, forming through allusion in the interstices of everyday communication. Placed at its limits, the possibility of non-recognition also affects the core and permeates the fabric of social interaction.[11]

Nevertheless, it is not possible to go back to a pre-linguistic form of immediate empathy. The only possible way of disciplining this "evil" dimension of language is by taking it to its logical conclusion. The power to objectify, torture and kill a fellow human animal, reducing it to pure inhumanity, is also the power to heal it and to create something new, contributing towards the formation of a community for the multitude to come.

3

In *The Violence of Financial Capitalism*[12] and *Capital and Language: From the New Economy to the War Economy*[13] Christian Marazzi adds a different and economic perspective to the argument for the contemporary relevance of the linguistic practices of human nature. We do not have space here to go into the details of the many descriptive points that are addressed in these two works. Marazzi is a thinker who always writes with the urgency of the historical moment and therefore in his books, for instance in the first part of *The Violence of Financial Capitalism*, we find some of the clearest and most effective accounts of certain economic and financial conjunctures (for example we find an enlightening and brilliant account of the 2008 crisis of subprime lending). We will focus here on some of the points that more clearly intersect with the issue of human nature's relevance to the contemporary form of capitalist accumulation.

In his first work translated into English, *Capital and Language: From the New Economy to the War Economy*, Christian Marazzi – within a debate on post-Fordism centered largely on the issue of the presence of language within the organization of the production process – describes a significant and definitive shift that occurred in the economy, stressing language's centrality to the spheres of credit and financial capital. This point allows him to produce an effect of resonance between the realm of production and the realm of finance, given that in post-Fordism both of them function by handling the very same elements: language and communication. His analysis of financial markets revolves around a crucial term: *convention*. In a refined and insightful argument, Marazzi stresses the self-referential nature of the stock markets and the fact that language therefore acquires a central role "not only as a vehicle for transmitting data and information, but also as a *creative force*."[14] The speculative nature of finance capital does not represent a pathology deviating from an alleged norm of correct measures, but rather a completely rational functioning proper to the stock markets that highlights the performative dimension of communication:

> Prices are the expression of the action of collective opinion, the individual investor does not react to information but to what he *believes* will be the reaction of the other investors in the face of that information. It follows that the values of securities listed on the stock exchange make reference to themselves and not to their underlying economic value. This is the self-referential nature of the markets.[15]

The language of the stock market therefore ceases to be a form of representation of an economic value external to it that would pertain to the sphere of material production; rather, it directly enacts this very value by means of the conventions of public opinion. At this point, Marazzi takes a brief but nonetheless crucial detour through the philosophy of language (the section entitled "Pathways of Language Analysis") in order to stress this non-communicational and intransitive dimension. Marazzi takes us from the biological foundations of language (Felice Cimatti) to feminist analysis of intrauterine language as an expression of our ontological being-in-relation (Luisa Muraro) in order to make us see the self-reflexivity and performativity proper to *any* enunciation. This is what constitutes a linguistic convention such as the one we see operating in the stock markets:

> A convention [...] is the fruit of a series of performative utterances, that is, utterances which do not describe a state of things but which immediately *produce* real facts. If we consider language to be not only an instrument used in institutional reality to *describe* facts, but also to *create them*, then in a world in which institutions like money, property, marriage, technologies, work itself, are all *linguistic* institutions, what molds our consciousness, language, becomes at the same time an instrument of production of those real facts. *Facts are created by speaking them.*[16]

Now, is it not the case that every utterance is based on a certain performative enunciation? What is so different about the post-Fordist shift to a language economy? Marazzi, by way of Benveniste, reminds us that the logic of enunciation usually makes its very efficacy dependent on the legitimacy of the person uttering it:

> There is a great difference if the person who says that the markets are prey to some

> irrational exuberance is Alan Greenspan or the present writer. The plot thickens when *even* Alan Greenspan, although speaking from the heights of his authority, no longer manages to modify the current state of affairs, for example when, announcing a reduction in interest rates, he fails to convince the community of investors of the real possibility of an economic recovery.[17]

What we see in the post-Fordist economy is the very crisis of this convention which Virno calls an *absolute performative*: a performative as such; the pure sign of a speaking-being without any other qualification (as in Virno, a pure enunciation without any enunciated content). The stock market crisis (Marazzi is referring in this book to the dot-com bubble of 2000) is therefore a crisis of *overproduction of self-referentiality*, an over-inflation of the self-referentiality of the convention itself. In typical workerist thinking every economic crisis is always a crisis caused by over-inflation: in the 1970s by salaries, in the 2000s by the overproduction of linguistic self-referentiality (which is also one of the theoretical points that Marazzi constantly addresses in his most recent work: how is an inflation crisis possible in a disinflationist economy?). But in this case "the crisis of a convention means the explosion of the *body* of the multitude, of the plurality of the individual differences which, once again, must face the, if you will, historical task of producing/electing a new convention."[18] The problem is that this crisis is also the crisis of "the multitude as 'natural antecedent,' its being by now a *historical result*."[19] What was naturally *always already* there (such as the pure potentiality of every utterance), *only now* takes *immediate* center stage. We are slowly seeing how in Marazzi's line of argument he is getting rid of every level of mediation in order to arrive at an idea of human nature "as such," "*sans phrase*," as a direct source of value, something which becomes even clearer in his most recent work.

In *The Violence of Financial Capitalism*, published in Italy almost ten years after *Capital and Language*, Marazzi addresses the genesis of the contemporary financial crisis with a critique of the most common and stereotypical interpretations of financial capital, which consider it to be a parasitic and purely redistributive form of accumulation (that supposedly hijacked the real, and of course morally good, industrial economy because of a lack of political control). His argument follows and deepens the line of thought of the first book, but with some significant and important shifts. The relation between the sphere of production and the sphere of financial capital is no longer understood in terms of an external analogy but in terms of a full and complete co-penetration of the two realms. Far from gaining autonomy from industrial production, finance is in fact an essential *dispositif* lying at the core of the contemporary form of capitalism, perfectly integrated into the *entirety* of the economic cycle. While a new demand to go "back to the real economy" hypocritically appeared after the crisis of 2008 as an urgent need to return to "making things" (in a revisited physiocratic anti-machine argument), the Italo-Swiss economist actually points us in the opposite direction. There is nothing perverse about financialization. What appears to be extremely problematic is rather the very dividing line that ideologically separates the real from the financial economy. Marazzi wants to underline the very connection between financialization and processes of *production* of value, without reducing financial economy to the sphere of *distribution*. As he says explicitly: "Financialization is not an unproductive/parasitic deviation of growing quotas of surplus value and collective saving, but rather the form of capital accumulation symmetrical with new processes of value production."[20] Financialization directly participates in strategies of creation of value. What, then, do those contemporary and finance-influenced processes of value production look like?

Marazzi proposes to consider the *extension* of the processes of value extraction way beyond what is traditionally defined as the mere production process (namely, the factory). In his view, production increasingly leads to the whole sphere of life itself progressively blurring the boundaries between labor-time and non-labor-time. We could say, for example, that production in post-Fordism enters into the sphere of circulation to the point of "transforming the consumer into a veritable producer of economic value."[21] As Marazzi (speaking through Marie-Anne Dujarier) puts it:

> *Coproduction*, where the individual is the coproducer of what he consumes, "is today at the heart of the strategies of public and private companies. *They put the consumer to work* in various phases of value creation. The consumer contributes to market creation, producing services, managing damage, and hazards, sorting litter, optimizing the fixed assets of suppliers and even administration."[22]

The typical example of this is the furniture megastore IKEA, where all the functions of individuating the code of the item, localizing it, transporting and assembling it, etc., are delegated to the consumer. It seems to be difficult to locate the dividing line between production and consumption.

But Marazzi goes a step further. This extension of the process of valorization is not only a hidden model of outsourcing, where the consumer is being put to work (like in a pure extension of the working day). What is at stake is also that *life* itself, in its most universal generality, becomes a source of value-production. Let us look at another example, so-called *crowd-sourcing*, that is, all those models, procedures and *dispositifs* aimed at extracting value from the crowd itself. The ability of Google, or web 2.0 websites like Facebook or Flickr, to attract and valorize the activity of users browsing these sites is another example of how the *core* of the production of value is being separated from the material production process. The success of those websites is in fact based on their ability to attract masses of users and make them share their whole world of social relations. What is the purpose of this? If it is true, as Marazzi and other postworkerist thinkers (like Maurizio Lazzarato or Carlo Vercellone) seem to believe, that in the digital economy it is possible to replicate commodities at decreasing costs, the very principle of economic scarcity seems to be relativized, to the point that – for example in purely linguistic or cognitive commodities – the very principle of scarcity is not operative at all (a material commodity cannot be possessed by two persons at the same time, while an immaterial one can be spread out over millions of copies without any loss). The fact that a website like Google is able to attract the attention of billions of users every day makes their platform have inestimable value, because the only true scarce commodity in cognitive capitalism is none other than attention itself.

While the cost of material goods decreases (because of the never-ending attacks against unions, outsourcing, etc., but also as a consequence of the innovation of the production process, technological advancement, etc.), the cost of attention increases:

> The *attention economy* is the product of the high rate of growth of access to information provoked by new technologies, in that in order to maintain or simply attract customers/consumers it is necessary to capture their attention. And this costs more and more all the time. It is a cost that increases as the unit cost of production decreases.[23]

This dimension of *capture* is the key to understanding what has been defined as *biocapitalism*, a form of production which is inextricably linked to the valorization of generic life itself. According to Marazzi, this form of extraction of value represents a new organic composition of capital (that is, the relation between constant and variable capital). Constant capital is constituted by all the linguistic machines, the procedures of value-capture, and all the *dispositifs* that follow citizen-laborers around every moment of their lives, while variable capital is constituted by the totality of sociality, relations, desires, language practices, etc., dispersed everywhere in society. The crucial point is that the basic means of post-Fordist production are therefore constituted by knowledge, know-how, protocols, forms of cooperation, *that do not have any material substance that is not their being part of the living body of the multitude*: therefore, they are constituted by their incorporation *into* life itself. Marazzi often uses terms like "extracting," "capturing," "sucking surplus-value," "harness and valorize," etc., that are highly symptomatic of what he defines as the "becoming-rent of the profit." In fact, this whole picture goes in the direction of portraying a contemporary financial capital that is entirely passive, no longer able to create surplus-value out of the living-labor of labor-power, and that is forced to go out of the factory gates to find value in the multiple forms of pure life of the multitude. Life becomes a *direct* force of economic valorization. Labor is reduced to a marginal and residual cost. What is striking in Marazzi's account of the "becoming-life-of-labor" is the underestimation of the entire concrete *dispositif* of the organization of production. If *life itself* (without being traversed by labor) has become the source of labor, why do the lives of millions of laborers around the world still retain such a horrible appearance?

In this regard, by removing the category of labor Marazzi's thesis goes resolutely beyond Marx (or against him). The problem is not only its overly optimistic tone and its reduction of all the political, social and economic mediations involved to a flat surface, where in the end every form of value production seems to be the same as every other. Even more problematic than this, as was the case with Virno, is Marazzi's general account of the category of life itself: by separating the category of labor from that of life, we will easily end up losing both of them. If *life as such* becomes the source of valorization, and capital does not have any role in organizing the production process (which, of course, includes the purchase of labor-power and its violent inclusion in the production process), capitalism will end up being a sort of collective hallucination. Why should every life not just live without caring about the fact that some parasite is trying to capture its value? Furthermore, if capitalism's participation in the concrete structuring of social relations is limited only to the position of the passive rentier, and if labor no longer has any importance as the source of value, what then is the differentiating principle of what we understand as life? How can different lives be different from each other if capitalism no longer structures and organizes their cooperation and their reproduction? To put it ironically: if capitalism is only a passive rentier, why are we still living in a capitalist world?

Labor, in Marxian terms, is not only an external disciplinary force inflicted on the worker, it is also the principle that defines the *qualitative* dimensions of production and reproduction. It is what defines the *purpose* of production (what to produce? Where? In what quantity? And the guiding principle is none other than the pure drive for abstract accumulation). Value is not only a quantitative measure that applies itself externally to the production process (and the fact that a production process like post-Fordism is allegedly based on non-scarce commodities reduces it to a passive and arbitrary measure); it is also the qualitative organization of production itself, and therefore the organizing principle of the reproduction of labor-power. Life had always been put to work long before capitalism, but so long as production is organized and managed by the extraction of surplus-value, *life* will always be internally differentiated by the capitalist mode of *re*production. The problem is that *life itself* will never encounter capitalism as an external or parasitic force, for the simple reason that as long as capitalism remains operative it will function as an internal differentiating principle (or attribute) of life itself. Is it not perhaps the case, as it was with Virno, that in order to encounter the pure genericity of life one should not dwell on post-Fordism, but rather step outside of capitalism?

notes

1 P. Virno, *Multitude: Between Innovation and Negation*, trans. Isabella Bertoletti, James Cascaito, and Andrea Casson (New York: Semiotext(e), 2008).

2 S. Freud, *Jokes and their Relation to the Unconscious*, trans. James Strachey (New York: Norton, 1963).

3 *Multitude* 81.

4 Ibid. 82.

5 *Jokes and their Relation to the Unconscious* 207.

6 *Multitude* 92.

7 Ibid. 73.

8 L. Wittgenstein, *Philosophical Investigations*, trans. G.E.M. Ascombe (Malden, MA: Blackwell, 2001).

9 *Multitude* 115.

10 Ibid. 119.

11 Ibid. 173.

12 C. Marazzi, *The Violence of Financial Capitalism*, trans. Kristina Lebedeva (New York: Semiotext(e), 2010).

13 Idem, *Capital and Language: From the New Economy to the War Economy*, trans. Gregory Conti (New York: Semiotext(e), 2008).

14 Ibid. 27.

15 Ibid. 26; my emphasis.

16 Ibid. 33.

17 Ibid. 34.

18 Ibid. 36.

19 Ibid.

20 *Violence of Financial Capitalism* 48.

21 Ibid. 50.

22 Ibid.

23 *Capital and Language* 66.

davide tarizzo

translated by alvise sforza tarabochia

THE UNTAMED ONTOLOGY

W*e, the moderns, live.*

But when have we started to live? When has modernity begun? When have we become ourselves? It is on these questions that Michel Foucault's most intense and vertiginous work, *The Order of Things*,[1] is grounded. Among the theses that he airs in this book, those that interest us the most are the following:

(a) Modernity does not begin, as is canonically believed, between the sixteenth and seventeenth centuries, with the birth of a new science and the revolution that René Descartes produced in philosophy. Modernity begins later, at the end of the Classical Age, more or less with Immanuel Kant. This is a thesis to which Foucault remains faithful until the end of his teaching, even when he makes the beginning of modernity coincide with that of the Enlightenment and sees in them the outline of what he calls an "ontology of actuality." In the end, this ontology of actuality (or of the present) is nothing other than an ontology of modernity.[2]

(b) But what is modern? Man is modern, or, better, humanity understood as a threshold of epistemic positivity, as a field of infinite investigations and inquiries, all aimed at answering the question: what is man? Before modernity, man did not exist, Foucault argues, in the sense that the question on the humanity of man was not being asked, nor was there any "analysis of finitude"[3] in which "man's being is always maintained, in relation to man himself, in a remoteness and a distance that constitute him."[4] Yet the profile of man understood as the subject and object of his own reflection, as an "identity separated from itself by a distance which, in one sense, is interior to it, but, in another, constitutes it,"[5] emerges precisely in this distance. All this circumscribes the horizon of modernity, or of our actuality, whose *critical* threshold is given, according to Foucault, by Kant's thought.[6] It is with Kant that the critical man comes into being, the man who interrogates the Self (the Same) of his intimate and critical humanity. It is with Kant that the anthropological "Fold" appears, the one by virtue of which man has to discover and give him-Self the law that will finally render him human.[7] Since Kant, this is what we understand as the autonomy of the modern man. The ontology of actuality is an ontology of autonomy.

(c) On the threshold of autonomy, which imposes "the ever-to-be-accomplished unveiling of the Same,"[8] or of the human Self, the three epistemic *a priori* of life, labour, and language surface as that by means of which the entire field of human knowledge is rearranged: biology, economics, and linguistic sciences are born in this way. The presence of biology, a natural science, is striking in a work that, from beginning to end, tries to reconstruct the archaeology of contemporary human sciences, with the purpose of picturing their invisible categorial framework. This is all the more remarkable because, only ten years later, Foucault proposes to find the secret cipher of modern power precisely in "biopowers" and "biopolitics." Modernity, Foucault argues, is the age in which a new form of power spreads, one that emancipates itself from the law of the sword, from the classical code of sovereignty, in order to take charge of life as such. Its motto is no longer *Take life or let live* but *Make live and let die*.[9] Power now targets life in itself, a life that needs to be cultivated, empowered, directed, and regulated. But what are the conditions of possibility of this modern apparatus of power? The most important one was identified by Foucault ten years earlier: it is the emergence of "life" as a historical-epistemic *a priori*, along with labour and language. Foucault's thesis is that, before modernity, life did not exist just as a science of life as such – biology did not exist. Conversely, once we enter modernity, we witness the simultaneous appearance of a knowledge about life and of a power over life, of biology and biopolitics, whose lethal connection we need now to grasp.

> Historians want to write histories of biology in the eighteenth century; but they do not realize that biology did not exist then, and that the pattern of knowledge that has been familiar to us for a hundred and fifty years is not valid for a previous period. And that, if biology was unknown, there was a very simple reason for it: that life itself did not exist. All that existed was living beings, which were viewed through a grid of knowledge constituted by natural history [...] Natural history, in the Classical period, cannot be established as biology. Up to the end of the eighteenth century, in fact, life does not exist: only living beings [...] The naturalist is the man concerned with the structure of the visible world and its denomination according to characters [...] not with life.[10]

This is the reason why we, the moderns, live. It is because modernity, the reign of autonomy, is the reign of life. We *are* no longer, rather, we *live*: it is therefore a new ontology that defines our condition. The knowledge about life (biology) and the power over life (biopolitics) are possible only against the background of a specific abstraction of life, an ontology of life in which we can recover the hidden roots of what Foucault will later name "ontology of actuality." But what is life, this modern ontological abstraction? The passages where Foucault discusses it are among the most heated and disordered of all his oeuvre. Nevertheless, we can draw three indications from them.

Life is a synthetic unity. In the Classical or pre-Modern Age, the naturalist casts his gaze on the living being only to give it a name and arrange it in the linear space of a representation, a classification. Only living beings existed, scattered in their multiplicity, which had to be brought back to a principle of order. The Classical Age is the reign of taxonomy. With the advent of modernity, on the contrary, the living being no longer acquires the meaning of "living" in relation to the other living beings and to its position among them, but in relation to the secret and synthetic unity that each of them *conceals*; "Multiplicity is apparent, unity is concealed" and henceforth living beings will be regarded as living "only because they are alive and on the basis of what they conceal."[11]

Life is a secret force. What living beings conceal, what makes them, strictly speaking, live, is life itself understood as a force. If the knowledge of the classical naturalist was a knowledge of the forms-of-life, the knowledge of those who observe nature is now a knowledge of the force-of-life that animates every living being and concentrates itself in a "sort of focal point of identity,"[12] which cannot be penetrated by our gaze. With modernity, we move from the *superficial* knowledge of the living beings to a *deep* knowledge of life itself; from a knowledge of visible forms to a knowledge of an invisible force. This force, "life," is that which folds every living being on itself; it is that which makes of each living being a Self, focused on the continuous endeavour to affirm and reinforce itself. Thus, while in the Classical Age "being was posited in the perpetually analysable space of representation," in the Modern Age "life withdraws into the enigma of a force inaccessible in its essence, apprehendable only in the efforts it makes here and there to manifest and maintain itself."[13]

Life is an obscure will of the Self. At this point, we leave the flat ontology of being and representation and enter what Foucault calls an "untamed ontology." The field of knowledge is no longer split between existing and non-existing things, divided by the border between being and non-being; it is now pervaded by something that neither is nor is not, but *wants to be* in every living being – life. Every form-of-life is reduced to the precarious and transient expression of a force-of-life, of a blind and eager will to exist, of a "mute and invisible violence that is devouring [...] in the darkness"[14] every living being, right after bringing it into existence. It is the dawn of a new ontology, that does not know the abyss between being and non-being, but dynamically insinuates itself in their tension. It is the dawn of modernity, the dawn of life, which replaces the old ontology of the one and the many with the untamed ontology of the Self.

> For life – and this is why it has a radical value in nineteenth-century thought – is at the same time the nucleus of being and of non-being: there is being only because there is life, and in that fundamental movement that dooms them to death, the scattered beings, stable for an instant, are formed, halt, hold life immobile – and in a sense kill it – but are then in turn destroyed by that inexhaustible force. The experience of life is thus posited as the most general law of beings, the revelation of that primitive force on the basis of which they are; it functions as an untamed ontology, one trying to express the indissociable being and non-being of all beings. But this ontology discloses not so much what gives beings their foundation as what bears them for an instant towards a precarious form and yet is already secretly sapping them from within in order to destroy them. In relation to life, beings are no more than transitory figures, and the being that they maintain, during the brief period of their existence, is no more than their presumption, their will to survive. And so, for knowledge, the being of things is an illusion, a veil that must be torn aside in order to reveal the mute and invisible violence that is devouring them in the darkness.[15]

Of which "knowledge" is Foucault speaking? What knowledge discloses itself in the horizon of life? Which discourses can take place in the context of a modern "untamed ontology"? Although his words might seem to recall, at first sight, some kind of metaphysics of life and the many vitalistic philosophies that came into being during the Modern Age, Foucault is actually convinced that the "untamed ontology" and its "synthetic notion" of life form the basis of a scientific, and not only metaphysical, knowledge. He is convinced that the ontological horizon of life includes within itself and presupposes as a possibility both a scientific and a metaphysical discourse.

Foucault shows more interest in the *scientific* side, which is the one he prevalently thinks about. In his opinion, the untamed ontology engenders a new, strictly modern, knowledge, entirely grounded on the synthetic notion of life. This knowledge, this new science of modernity, is *biology* – which he compares and contrasts with another modern knowledge, *economics*. Foucault is talking about biology when he says that "a system of thought is being formed that is opposed in almost all its terms to the system that was linked to the formation of an economic historicity." The latter was grounded on the "triple theory of irreducible needs, the objectivity of labour, and the end of history," while biology is, instead, a thought

> in which individuality, with its forms, limits, and needs, is no more than a precarious moment, doomed to destruction, forming first and last a simple obstacle that must be removed from the path of that annihilation;

> a system of thought in which the objectivity of things is mere appearance, a chimera of the perceptions, an illusion that must be dissipated and returned to the pure will, without phenomenon, that brought those things into being and maintained them there for an instant; lastly, a system of thought for which the recommencement of life, its incessant resumptions, and its stubbornness, preclude the possibility of imposing a limit of duration upon it.[16]

In this passage, it is possible to recover a further connotation of the synthetic notion of life which, for Foucault, forms the basis of modern biology: life, he says, is a "will without phenomenon." We still need to see up to which point such an acceptation of life may be regarded as scientific.

The *metaphysical* side constantly remains in the background, always ready to re-emerge in the framework of the untamed ontology of modernity. According to Foucault, this ontology contains in itself the seeds of both a science and a metaphysics of life. Biology and vitalism share the same conditions of possibility. And the appearance of the one is simultaneous to the (re)appearance of the other.

> It is this transition from the taxonomic to the synthetic notion of life which is indicated, in the chronology of ideas and sciences, by the recrudescence, in the early nineteenth century, of vitalist themes. From the archaeological point of view, what is being established at this particular moment is the conditions of possibility of a *biology*.[17]

From Foucault's perspective, this is a fundamental hypothesis: the archaeology of modern vitalism and the archaeology of modern biology are *the same archaeology*, because they lead us to the same notion of life, to the same, deep, semantics of an "untamed force." The validity of Foucault's archaeological investigation is confirmed precisely by the identical epistemic *a priori* – that is, life – of two discourses that have different registers and formulations. What kind of relationship is then established between a science of life and a metaphysics of life, if both are rooted in the depths of the same *a priori*? If, according to both, the ultimate secret of life is the "perpetual devouring of life by life"?[18] Foucault's answer is evasive and takes the shape of a further question that eludes the original one, showing us the limits of his own investigation.

The answer/question is the following:

> Must we admit that from now on each form of [epistemic] positivity will have the "philosophy" that suits it? Economics, that of a labour stamped with the sign of need, but with the eventual promise of the great reward of time? Biology, that of a life marked by the continuity that forms beings only in order to dissolve them again, and so finds itself emancipated from all the limitations of History?[19]

Foucault does not answer the question in this context, nor will he ever answer it elsewhere. Yet it is possible to guess what he is insinuating. Just as economics, as a modern science, is always inclined to turn into a philosophy of history with which it would share the same epistemic *a priori*, so biology is always inclined to turn into a philosophy of life with which it would share the same deep and "untamed" semantics. In other words, there is no gap between a science and a metaphysics of labour, just as there is no interruption between a science and a metaphysics of life. But why is it that Foucault does not go beyond this point?

One of the reasons why he does not go beyond it, the most superficial, is that he does not need to do it in the context of an "archaeology of the human sciences." To fulfil this task, he only needs to punctuate the discontinuity between the epistemic paradigms of the Classical and the Modern Age. He does not need to detail the map of the reciprocal intersections that are produced, within the borders of the same epistemic paradigm, modern or classical, between different regimes of enunciation – in this case, the philosophical and the scientific. An archaeological analysis is differential and discontinuist, which is why it does not aim at reconstructing from scratch epistemic organisms, articulated bodies of knowledge; this kind of analysis rather intends to emphasise gaps and intervals. It does not have to deal with what is full, but only with voids, from whose margins one can detect leaps and fractures. It is true that each individual epistemic paradigm seems to be crossed by tensors that consolidate its profile here and there. Foucault himself, for instance, seems to be perfectly aware of the fact that modern man, vexed by a desire that manifests the "perpetual and fundamental situation of scarcity"[20] and fixes his constitutive distance from the Self, *not coincidentally* emerges together with the modern notion of life. This very awareness sheds light on some of Foucault's most suggestive pages, in particular those in which the themes of the will and of life seem to pass into each other, as inflections of a single untamed force that ends up coinciding with the peculiar "freedom" of modern man. A passage such as the following is symptomatic in that it opens and closes any possible question in the sudden flash of an intuition:

> The violence and the endless effort of life, the hidden energy of needs, were all to escape from the mode of being of representation. And representation itself was to be paralleled, limited, circumscribed, mocked perhaps, but in any case regulated from the outside, by the enormous thrust of a freedom, a desire, or a will, posited as the metaphysical converse of consciousness. Something like a will or a force was to arise in the modern experience – constituting it perhaps, but in any case indicating that the Classical age was now over.[21]

Nevertheless, there is another reason why Foucault chooses not to proceed in the direction that we are trying to indicate here. This second reason is, in all probability, a reasonable caution. After all, if one gets to the point of affirming that there is no interruption between a science and a metaphysics of labour, we all know where this leads us to, namely, the first great metaphysical sequence of modernity that, from Georg W.F. Hegel, brings us directly to Karl Marx. What is more, the science of economics could not boast an incontrovertible scientificity. This holds good today as much as it did then. What should we say of the second sequence? To affirm that between a science and a metaphysics of life there is no interruption means venturing into a minefield. It means openly confronting one of the most established scientific dogmas and one of the most authoritative scientific names of modernity. And, quite possibly, in an already daring work such as *The Order of Things*, Foucault does not want to venture that far.

Those who have read the book know the stratagem that Foucault devises. He chooses not to mention the true founder of modern biology, in order to retrieve the conditions of possibility of this modern knowledge in an author that precedes him. Foucault maintains that the modern synthetic notion of life, with its ambiguities and deep semantics that cross the limits of scientific enunciation and trespass into "philosophy," originates with Georges Cuvier. It is in Cuvier that life sparkles, beyond the living beings, in the constitutive gap between each living being and its own Self. It is in Cuvier and not, for instance, in Jean-Baptiste Lamarck.

> Though it is true that Lamarck, by the influence of a retrospective illusion, has been overestimated at the expense of Cuvier, though it is true that there is little awareness of the fact that "life" reached the threshold of its positivity for the first time with the *Leçons d'anatomie comparée*, there is nevertheless at least a diffused consciousness of the fact that Western culture began, from that moment onward, to look at the world of living beings with new eyes.[22]

But why is it that Cuvier is so important after all? To put it in Foucault's words, because Cuvier is the condition of possibility for Charles Darwin. Because Cuvier synthesises a notion of life that becomes, in a short time, the centre not only of Lamarck's but also of Darwin's theory of evolution. "Evolutionism is a biological theory, of which the condition of possibility was a biology without evolution – that of Cuvier."[23] Hence, the untamed ontology, that which, according to Foucault, is modern life's pulsating ontology, includes – not only, but *also* – Darwin's evolutionary ontology, the ontology of natural selection, of which Cuvier's biology is a metonymy.[24] This is the bridge, which could be inferred from Foucault's veiled hints, between a science and a metaphysics of life. This is the intuition before which Foucault preferred to stop.

Yet Foucault knew what we all know, namely, that in modern biology "life should be defined by the possession of those properties which are needed to ensure the evolution by natural selection."[25] He knew it as much as he knew that "the doctrine of natural selection" is not only a doctrine that "unifies the whole biology" but also, in no uncertain terms, "the only possible" doctrine that can fulfil this task.[26] Likewise, Foucault knew that this doctrine penetrated the newly opened gap between the living being and itself, between "the word and the thing," which were still perfectly interwoven in the imaginative naturalistic inquiries of the Classical Age.[27] He also knew that the Darwinian theory could discern in any form-of-life and in any living being the "outcome and temporary stopping-place of a continuous dynamism which itself must be termed 'life'"[28] – a life that, understood in this way, eventually assumes a meaning that invisibly brings it beyond the borders of scientific enunciation. Perhaps Foucault feared the misunderstandings and the frowning silence caused, a few years later, by Karl Popper's famous paper "Darwinism as a Metaphysical Research Programme."[29] Popper's thesis was that the theory of natural selection worked on the grounds of a "situational logic," and hence could be valid only in a specific hypothetical context in which a pre-established notion of "life" was taken for granted. This did not mean that Popper completely denied the scientific validity of Darwinism. Rather, he meant to show its metaphysical side, which, for him, remained indiscernible, for better or for worse, from its properly scientific side. At any rate, his critique was not, and did not intend to be, a *refutation* of Darwinism.[30] In the same years, Foucault racked his brain with an analogous idea, even if he did it with greater caution. This is witnessed by a television dialogue with Noam Chomsky during which the French intellectual goes as far as claiming that, in modern biology, "life" does not function as a normal scientific concept, but as an "epistemological indicator," as a sort of scientific meta-concept, deprived of a real referent, yet still fundamental to "designate, to delimit, and to situate"[31] biological discourse. What did Foucault mean when he equated life with an "epistemological indicator," that is to say, a concept deprived of any connection to reality, whose only function would be that of circumscribing an epistemological field? And how should we interpret the last pages of a history of biology published in the same years by a well-known scientist (whom Foucault highly praised), in which it is claimed that "life" is a metaphysical entity?[32]

It would be wrong to read in these statements a general invalidation of the premises and procedures of inquiry of modern biology, as if it were possible to reduce them to the smoke of a volatile *philosophical theory*. Foucault and Popper did not have this in mind. They could not have agreed with what, for instance, George Lakoff claims.

> Philosophy is most powerful when it is invisible. Over the course of centuries philosophical theories may become so engrained in our culture and our intellectual life that we don't even recognize them as theories; they take on the cast of self-evident truth, part of the intellectual landscape that serves as a background for theorizing. Such virtually invisible philosophical theories are often harmless. But when they are false and become widely accepted within important academic disciplines, invisible philosophical theories can stand in the way of scientific investigation. Because they are invisible, they are neither questioned nor taken into account.[33]

The situation delineated here is one in which, almost like in a laboratory, there would exist theoretical compartments that are watertight, in the specificity, the compartment of philosophical theories, scientifically false and unfounded, and the compartment of scientific theories, true and well-founded. It goes without saying that, in a situation like this, to demonstrate that a theory is philosophical rather than scientific would mean, *ipso facto*, demonstrating that the theory is false and betrays its initial aspirations. But on closer inspection this is not the case, since it is only rarely that we are able to outline clear-cut and impassable demarcation lines between enunciative registers that, nonetheless, remain separated in many aspects. Words continuously migrate from one region of our epistemic horizon to another and, in doing so, they bring along with them invisible effects of meaning. This does not entail that they bring along with them entire *saturated*, *coherent* and *unitary* theories. Even if they belong to different discursive regimes, enunciations rather seem to contaminate and stain each other, here and there, obfuscating our futile, albeit licit, dreams of integral purity. In other terms, the words we utter, wherever this may happen, are soaked in history and, for this reason, blurred.

This problem is better addressed by Charles Taylor in a work that also deals, in a remarkably original way, with the thorny issue of modernity. Taylor's hypothesis is that, rather than philosophical theories, there are *invisible ontologies*, buried in the generalised and diffused "frameworks" through which we orient ourselves in the world.[34] Taylor says that these frameworks are historical, they change in time (not unlike Foucault's epistemic paradigms) and impose precise "qualitative distinctions," drawing the boundaries of alternative moral ontologies. From this perspective, what is the role of philosophy? It is not that of inventing and championing, at each turn, a new invisible ontology, a new moral ontology, almost as if it were a new theory. Rather, its role is more limited and humble, that is, it amounts to unearthing and bringing to light, at each turn, the deep semantics of certain nuclear and central concepts around which one "framework" or the other revolves. The owl of philosophy takes to flight at dusk, when all is said and done: this was Hegel's warning. Which does not mean that it cannot cast its gaze afar, to the point of seizing the margins of an entire horizon, which would otherwise be doomed to remain in the shadows; this does not mean either that it cannot, in the end, turn its own horizon – its own time – against itself, making us lean out of the vertiginous ridge of its contingency.

The untamed ontology of which Foucault speaks is, without doubt, an invisible ontology in Taylor's sense, even if it cannot be reduced to the perimeter of moral values. It is by all means an ontology that has strong moral and, before that, political implications, but it is also an ontology that has heavy epistemic consequences. It is centred on a synthetic notion of "life," an abstraction of "life," which is, on the one hand, the condition of possibility of biopolitics and perhaps even of a religion of "life" which other authors have investigated.[35] On the other, it is the condition of possibility of both a metaphysics of "life," whose echoes are unprecedented, and a science of "life" that operates, without knowing it, within the same horizon. This notion, this abstraction, *neither exhausts nor saturates* the discursivity of "life" in its different inflections; rather, it *orients and conditions* it. Modern "life" has its own *density*, wherever we use the term, be it in a philosophical or scientific, moral or political context. It must be stressed that this density does not reabsorb or recapitulate within itself every discourse on "life," but rather puts forward and irradiates a secret force that *bends* these discourses in a certain direction. And it makes sure that, each time that we speak about "life," it is us, the moderns, who speak.

Our aim is to decipher, in Foucault's wake, the semantics of modern "life," in order to complete or at least to broaden the framework of the untamed ontology which he sketched only in part. In doing so, we will set forth from a thesis whose meaning we will have to clarify: *the metaphysical threshold of modernity is autonomy*. Beginning from this thesis, it should be possible to link what Foucault was never able to link: the semantics of the autonomous will (which, for instance, forms the basis of modern juridical notions) and the semantics of autonomous life, that is to say, the two sides of the current age. Beyond the threshold of autonomy, the will and life appear as the two complementary sides of our time, as the two surfaces of meaning of modernity which pass into each other, as if they were on a Möbius strip. The will and life appear as the two sides of this strip, our present, that permeate each other. But we, the moderns, are neither of them. We are only the fold that enables the strip of the present to chase itself and coil around a void. We, the moderns, are not.

We live and we will. "The irony of this deployment is in having us believe that our 'liberation' is in the balance."[36] It makes us believe that what is at stake is a freedom, our freedom, which would render us human, while inside of us rages a battle that is not that human. "We do will ourselves . . ."[37]

notes

1 M. Foucault, *The Order of Things: An Archeology of the Human Sciences*, trans. Alan Sheridan (London and New York: Routledge, 2002).

2 Idem, "What is Enlightenment?," trans. Catherine Porter, in *The Foucault Reader*, ed. P. Rabinow (New York: Pantheon, 1984).

3 *The Order of Things* 366.

4 Ibid. 367.

5 Ibid. 370.

6 M. Fimiani, *Foucault* e *Kant. Critica clinica etica* (Naples: Città del Sole, 1997).

7 *The Order of Things* 371ff.

8 Ibid. 370.

9 M. Foucault, *Society Must be Defended: Courses at the Collège de France 1975–76*, trans. David Macey (New York: Picador, 2003) 241.

10 *The Order of Things* 139, 175, 176.

11 Ibid. 291.

12 Ibid.

13 Ibid. 297.

14 Ibid. 303.

15 Ibid.

16 Ibid. 304.

17 Ibid. 293.

18 Ibid. 302.

19 Ibid. 304.

20 Ibid. 279.

21 Ibid. 227.

22 Ibid. 306.

23 Ibid. 320.

24 M. Foucault, "La Situation de Cuvier dans l'histoire de la biologie" [1970] in *Dits et écrits* (Paris: Gallimard, 1994) II: 595–618.

25 J. Maynard Smith, *The Problems of Biology* (Oxford: Oxford UP, 1986) 7.

26 J. Gayon, *Darwinism's Struggle for Survival: Heredity and the Hypothesis of Natural Selection*, trans. Matthew Cobb (Cambridge: Cambridge UP, 1998) 184.

27 J. Roger, *Les Sciences de la vie dans la pensée française du XVIII siècle. La Génération des animaux de Descartes à l'Encyclopédie* (Paris: Albin Michel, 1993) 53ff.

28 H. Jonas, *The Phenomenon of Life: Toward a Philosophical Biology* (Evanston, IL: Northwestern UP, 2001) 45.

29 K. Popper, "Darwinism as a Metaphysical Research Programme" in *Philosophy of Science: Contemporary Readings*, eds. Y. Balashov and A. Rosenberg (London and New York: Routledge, 2002) 302–20.

30 Popper, "Darwinism as a Metaphysical Research Programme."

31 M. Foucault and N. Chomsky, "Human Nature: Justice vs. Power" [1971] in *The Chomsky–Foucault Debate: On Human Nature*, by M. Foucault and N. Chomsky (New York and London: New, 2006) 6.

32 F. Jacob, *The Logic of Life: A History of Heredity*, trans. Betty E. Spillman (Princeton: Princeton UP, 1993).

33 G. Lakoff, "Cognitive Semantics" in *Meaning and Mental Representations*, eds. U. Eco, M. Santambrogio, and P. Violi (Bloomington and Indianapolis: Indiana UP, 1988) 122.

34 C. Taylor, *Sources of the Self: The Making of Modern Identity* (Cambridge, MA: Harvard UP, 1989).

35 B. Duden, *Disembodying Women: Perspectives on Pregnancy and the Unborn*, trans. Lee Hoinacki (Cambridge, MA: Harvard UP, 1993); R.M. Dworkin, *Life's Dominion: An Argument about Abortion and Euthanasia* (New York: HarperCollins, 1993).

36 M. Foucault, *The History of Sexuality, Volume I: The Will to Knowledge*, trans. Robert Hurley (New York: Vintage, 1990) 159.

37 M. Heidegger, "The Self-Assertion of the German University," trans. Karsten Harries, *Review of Metaphysics* 38 (1985): 480.

paolo virno

translated by lorenzo chiesa

THE ANTHROPOLOGICAL MEANING OF INFINITE REGRESSION

introduction

The cornerstone of a materialist anthropology that remains in tune with the theory of evolution and is nevertheless capable of accounting for the functioning of political institutions or the meaning of religious rituals consists of three fundamental logical categories: negation, modality of possibility, and infinite regression. The human being is specifically the animal that can introduce the little word "not" into any proposition; that is to say, the one that always knows how things are *not*. The human being is specifically the animal that, by using the phrase "it is possible that," shows a lack of orientation in the environment and, at the same time, a dexterity in compensating for it through the elaboration of behaviours which are not fixed in advance. The human being is specifically the animal whose thought and passions collapse at times into an interminable backwards march, sanctioned by the formula "and so on, to infinity."

In addition to exhibiting the salient characteristics of our form of life, the three categories I have just mentioned at the same time constitute, as a whole, the *logical basis of metaphysics*. Without negation, modality of possibility, and infinite regression, issues such as "being *qua* being," the search for a foundation or "first principle," the relation of the One to the Many, the couple true/false, the notion of limit, the articulation of historical time, and so on, would not even be conceivable. The metaphysical tradition only drew – different and even opposite – consequences from the functioning of the "not," the resources of the "it is possible that," the looming of the "and so on" without outcome. There is no ontology or theory of subjectivity that does not issue from a specific interpretation of these elementary apparatuses of verbal thought. A history of philosophy that focused only on reconstructing the variable modes in which a single and identical logical-linguistic score has been orchestrated would be trustworthy and exhaustive.

Everything seems to point to the fact that metaphysics, with its characteristic repertoire of non-empirical problems, is a *natural* tendency of our species, one that could be explained by starting from some decisive anthropological prerequisites. Yet metaphysics, as such, says nothing about these prerequisites. Things are different when it comes to its logical basis. Indeed, such a basis coincides with the distinctive traits of *Homo sapiens*. If, on the one hand, negation, modality of possibility, and infinite regression outline the contours of *philosophia prima* (allowing us to ask such questions as: "Why is there something rather than nothing?"), on the other, they amount to the syntactic equivalent of significant phylogenetic matters of fact (for instance, the retention of infantile characteristics into adulthood and a related poverty of innate inhibitions). The logical basis of metaphysics, unlike what has at each turn been built on it, corresponds to a set of adaptive functions, distils the ways in which the linguistic animal carries out cognitive and operative tasks on which its survival depends, and stresses the framework of emotions and affects that mark its existence.

In what follows, I will say nothing, or very little, about negation and the modality of possibility. I would like to dedicate an independent study to the prerogatives of the sign "not," and its centrality in the material and sentimental existence of the human animal. I have discussed the category of the possible, at times remaining trapped in it, in all that I have written in the last twenty-five years:[1] I don't have much to add, and I don't find it polite to repeat myself. Here, in order to show how pressing and intricate the relation between logic and anthropology is, I will deal with a single topic: the meaning of infinite regression in the natural history of *Homo sapiens*. Far from being a bizarre and marginal eventuality, or a matter that should concern only professional logicians, the interminable "and so on" closely concerns every kind of cognition, practical behaviour, and affect. A child is already familiar with it when he asks what the reason is for a certain occurrence, and then the reason for this reason, and then also the reason for the second and more fundamental reason etcetera, thus giving rise to a vertiginous ascending hierarchy of "because." Infinite regression is a sort of *refrain*, familiar and uncanny at the same time, that accompanies, and to a certain extent conditions, every experience. If we do not pay due attention to the pervasiveness of this logical-linguistic phenomenon, we understand little of the ways in which our species adapts (or does *not* adapt) to its vital context, and, similarly, of the social and political conflicts that fill its history.

Infinite regression, which in logic marks the failure or incompleteness of a demonstration, is first and foremost a permanent possibility which the existence of the higher primate called *Homo sapiens* is exposed to. What, precisely, does this possibility amount to? For the time being an unassuming definition should suffice. It is correct to speak of infinite regression whenever the solution to a problem, no matter whether it is practical or cognitive, only proposes again at a more abstract level the very problem which it seemed to have just unravelled. Or also: whenever the overcoming of a limit has its confirmation as an inevitable result.

The metaphysical tradition offers many examples of it. The unstoppable backwards march that, according to the late Plato,[2] seems to afflict the relation between a sensible being and the corresponding universal idea should be regarded as almost proverbial. If we affirm that a singular empirical man participates in the idea of Man in the same way in which a copy imitates its model, we also need to specify what the two terms have in common: their reciprocal similarity presupposes, as a matter of fact, the existence of a shared unit of measure. This unit of measure can only be a model (that is, an idea) of a *higher level*, to which both the empirical man and the first idea of Man conform. We therefore return to what Aristotle calls, with a succinct formula, the "third man."[3] He is the third, but not the last. It is evident that even the meta-idea of Man must share a model (or idea) with the two entities that participate in it; and so on without end. Here is another example. According to Kant, whenever reason has to confront a conditioned phenomenon, it postulates that "the entire sum of conditions, and consequently the absolutely unconditioned (through which alone the conditioned has been possible) is also given."[4] Having said this, he discovers that, due to the fact that each premise is in turn a consequence, we face a regressive series in which each stage (the condition) refers back to the one that precedes it (the condition of the condition): "This regress is to be entitled infinite,"[5] in so far as each provisional completion reproduces the incompleteness which it had to provide for. In another symptomatic case, the American philosopher

Josiah Royce[6] has shown how the attempt to represent the mental image that a person has of his mind sets off an increasingly complex spiral. The image we create of our mind is itself a mental state which we need to account for. From this follows that the image of our mind, in order to be credible, must also be the image of the image of our mind. But the new and more comprehensive image is also a mental state, and so on.

It would not be difficult to come up with a plethora of further examples. Those I have chosen are nevertheless not entirely accidental. As we shall see more closely below, they refer, in a way that is less indirect than it might seem, to some typical prerogatives of the human animal. From an anthropological standpoint, the "third man" conceals the gap between the individual and the species, that is to say, the partial incommensurability of two terms which, at the same time, are complementary and inextricable. On the other hand, the ascending series of conditions that makes a given phenomenon possible points to a high degree of environmental lack of adaptation; or, in other words, *Homo sapiens*'s ability to sense and thematise the limits of any particular environmental configuration in which it happens to operate. As for the endless spiral caused by the attempt to elaborate a mental image of our mind, it illustrates well the perennially deficient structure of self-reflection.

compulsion to repeat

We have just said that infinite regression arises when the solution to a problem causes the re-emergence of the very same problem; when the overcoming of a limit is such that it entails its inexorable re-affirmation. Although they are certainly necessary, these conditions are nevertheless not sufficient. If we kept only to them, at most we would have to do with a circle, and, thus, with a process in which the result leads back to the starting point and the conclusion generates its own premise. But a circle is something different from an authentic regression. We need something else in order for an incessant movement backward (or forward – the two spatial metaphors, regression and progression, are to a large extent interchangeable) to effectively be unleashed. We need to adopt the same solution *each time anew* with reference to the problem that, on the other hand, it is doomed to pose again; we need to overcome the limit, which is nonetheless restated by its own overcoming, *always again* in the same way. To what is due this constant reiteration that alone allows us to really focus on the logical and anthropological figure of infinite regression?

We could suppose that, although it is as such a typically linguistic phenomenon, regression is embedded in the sub-symbolic field of human existence; that is to say, that it originates from a pre-verbal drive and feeds on it. In all truth, we could suppose that the "and so on" constitutes a refined, and in some aspects extreme, manifestation of the *compulsion to repeat* studied by Freud.

Let us recall the salient traits of this tyrannical propensity to the "one more time." The meticulous recurrence of the same gestures is a source of pleasure in infantile games but also the painful symptom of an unresolved past in the case of an obsessional neurotic. Whatever the forms and emotional tonalities with which it appears, the resolute aspiration to an eternal return of the same suggests, according to Freud, a general truth about the nature and function of our drives. Contrary to what is usually believed, the latter do not promote change and development but rather express "the conservative nature of living substance,"[7] their resistance to an excess of stimuli. This kind of resistance, far from being static, has the appearance of an indefatigable march backwards. The compulsion to repeat embodies "an urge inherent in organic life to restore an earlier state of things which the living entity has been obliged to abandon under the pressure of external disturbing forces."[8] Therefore, the boundless reiteration of the same behaviour aims at the restoration of "an old state of things, an initial state from which the living entity has at one time or other departed and to which it is striving to return by the circuitous paths along which its development leads."[9]

We should not lose sight of the features that make the forced repetition of which Freud speaks similar to the logic of the "and so on." Infinite regression itself stages a sort of eternal return of the same; dwelling on the examples I have already given, it is undeniable that we always have to attend again, until the end of time, to the idea of Man (in the case examined by Plato), or to the mental image of our mind (if we fall into the self-reflexive vortex described by Royce). Furthermore, regression also strives to go back to "an old state of things, an initial state": as a matter of fact, its unreachable goal amounts to a condition which is so primordial that it cannot be conditioned. This absolute presupposition coincides in many regards with a state of prehistoric balance, or better, with the threshold from which anthropogenesis began.

Despite the points of contact I have just indicated, the analogy between the compulsion to repeat and infinite regression is nevertheless completely mistaken. And this is so for an obvious reason, namely, in infantile games and the behaviour of the neurotic every new recurrence of the same gesture is placed *side by side* with the previous one, and thus located on the same plane, in such a way that no perspectival order can be established, not to mention a relation of subordination. The iteration of the drives is merely paratactical, that is, it is based on the simple conjunction of equipollent episodes: I act like this *and* again like this *and* again like this. The last "this" in the order of time entirely preserves the immediacy of the first; $n+1$ does not presuppose n, nor does it derive in some way from it. The further repetition is indiscernible from those which have already taken place and it could substitute them. On the other hand, in infinite regression, every successive step constitutes a development with regard to those that precede it; it inaugurates a more complex and comprehensive (epistemic and operational) level that subordinates the levels that have taken shape up to that instant. Just to make things clear: the nth image that a person has of his own mind is not comparable with those that precede it in so far as it includes and accounts for them; in the same way, the nth idea of Man raises one step above all others in so far as it provides them with a common unit of measure. Infinite regression overthrows the paratactical conjunction since it is rather characterised by a stringent hierarchical stratification. In logical regression alone, not in the compulsion to repeat, every term is generated from the one that precedes it in a kind of architectonical concatenation or perspectival vanishing point; in it alone does the overcoming of the limit *imply* the enlarged reproduction of the same limit – one that is able to obtain a level of higher generality.

Consequently, we have to ask: what is the origin of this hierarchical stratification on the basis of which the repetition of an identical movement fuels an enlarging spiral? It is again useless to think that we can find a satisfying answer in our congenital drives. We should rather pay attention to the apparatus that governs every aspect of a symbolic experience that is properly human, namely, syntax.

recursion

Infinite regression is a very particular manifestation of *recursion*, that is to say, the precondition that, according to Chomsky,[10] is most suited to distinguishing verbal language from the communicative codes of other animals and from any kind of sub-symbolic expressivity. Let us limit ourselves to the most elementary and intuitive meaning of this concept, leaving aside its complicated logical and mathematical developments (that is, its many connections with Gödel's theorems on the incompleteness of formal systems[11]). We call "recursive" the rule or procedure that is iteratively applied to the result of one of its previous applications. Here is a standard example. Any speaker, having inserted a subordinated clause within a main clause, can repeat the same operation in regards to what he has just obtained, inserting a new clause within the subordinated clause, and then another one within the second subordinated clause, and so on: "I think that / you know whether / Maria hopes that / Mario left." Let us also consider the iterative use of negation, dwelling on its most remarkable cases: I am allowed to append the symbol "not" before a clause that already contains it ("I *cannot* not marry you"), and then use it again to disavow the double negation I have just produced ("It is *not* the case that I cannot not marry you"), and so on. Furthermore, it is worth noting that syntactic recursion alone enables the human animal to master the progression of natural numbers: what else are we doing when we move from 10 to 11 and then to 12 etc. but adding a unit to a number that was in turn derived through the addition of a unit?

Building on a suggestive statement made by Humboldt, Chomsky has claimed that recursion guarantees the speaker the possibility of making "an infinite use of finite means." The repeated application of the same procedure to that which it has already produced explains the *innovative* character of verbal language, its independence from environmental influences and psychological inclinations.[12] The "one more time" inherent to syntactic recursion, unlike the one subjected to the compulsion to repeat investigated by Freud, does not at all produce a hypnotic paralysis; on the contrary, the iteration gives rise to an original state of affairs, encourages variation, and strengthens the capacity to accede to unexpected situations with ductility. Therefore, regression without an outcome is a specific version of the "infinite use of finite means," a collateral effect of the constitutive creativity of verbal language. The usual examples are instructive. The procedure "forming images of one's mind" is applied recursively to the image that I form of my mind; another idea of Man, capable of measuring the first unit of measurement, is applied recursively to the idea of Man as a unit of measurement of our species. The most significant case, from which all others follow, is, however, the ascending hierarchy of metalanguages; as a matter of fact, here recursion produces a regression without outcome precisely in a field, verbal discourse, which constitutes its natural place of residence. I can speak of the use of certain words: "The sentence 'the cat is on the mat' is correct"; but I can also speak (indeed, recursively) of the use of the words with which I have spoken of words: "A sentence such as 'The sentence "the cat is on the mat" is correct' is somehow pedantic"; and so on, to infinity. At each turn, the higher and more comprehensive metalanguage, just as every successive image of the mind and every further idea of Man, is really something *new*. Precisely in so far as it is based on recursion, infinite regression allows us in a way to reach "a state of things which has never yet been attained."[13] The change and development opposed by subsymbolic drives through the compulsion to repeat are no longer ascribable to "external disturbing and diverting influences"[14] but depend to a large extent on the very functioning of verbal language.

Hierarchical stratification and architectonical concatenation, which are completely absent from the compulsion to repeat, attest beyond any possible doubt to the absolute role played by syntactic recursion in infinite regression. Recursion is creative. Having its fundamental impulse in it, regression is itself creative, or at least could seem to be so. Overcoming *always anew* the limit means re-establishing it again, yet at a level of increased generality, and thus as an *always new* limit. However, we are talking about a "creativity" that is rather bizarre for it hinges on the monotonous re-proposition of an identical nucleus of experience. Monotony is not really affected by the fact that such a nucleus of experience, for instance a cognitive or practical problem, presents itself again in a more and more abstract form. In any case, what prevails is a *trompe l'œil* effect: the incessant innovation turns into a tautology, and the tautology takes the form of an innovative movement. What is this inextricable mixture of identity and difference due to?

The answer is intuitive. Inasmuch as, in infinite regression, recursion is grafted onto a logical circle, that is, an inference in which the conclusion lies at the foundation of the premise, it only offers a simulacrum of change and development. In this case, the syntactic apparatus based on the iterative application of the same procedure does not enable us to elaborate new semantic contents. *Syntax parts ways with semantics*: the former exhibits in no uncertain terms its transformative power, but this power goes round in circles, manifesting itself simply in the inexhaustible proliferation of the hierarchical levels; on the other hand, the latter appears to atrophy for it is compelled to keep up the pace. Infinite regression is an anthropological vestige of extraordinary importance even just for the fact that it documents the intertwining, which is typical of our species, between the irreversibility of the processes of development and the eternal return of the same, line and circle, innovation and "one more time." The linguistic animal is defined by the coexistence, or rather by the reciprocal implication, of these two possibilities. Regression indicates the way in which each of them runs through the other: it is precisely the irreversibility that prepares the return of the same, and the return of the same that paves anew the way to irreversibility.

syntax and drives

Having said this, I would like to advance a more circumscribed hypothesis. Infinite regression relies exclusively on the recursion of verbal language. It depends entirely on a syntactic precondition, certainly not on a motion of the drives. But its genetic independence from the drives and, in particular, from the compulsion to repeat, does not in the least entail an absence of relation with them. It is legitimate to conjecture a hybridisation of phenomena that are endowed with reciprocal autonomy; a hybridisation or, in the case at hand, a feedback. This is the crucial point: recursion, in the specific instances in which it stimulates infinite regression, connects with the compulsion to repeat and remoulds it from end to end. It projects onto the symbolic plane "an urge [. . .] to restore an earlier state of things" translating it into the forms of verbal thought. This further statement only reinforces the distinction I have previously drawn, rather than attenuating it. Infinite regression does not derive in any possible way from sub-symbolic drives yet, having its only foundation in syntax, *retroacts* on such drives and modifies their status and dynamics at the root. Recursion, to which the flight backwards of metalanguages and the ascending hierarchy of the mental images of one's mind are due, does not protract the compulsion to repeat, but replaces it with a functional equivalent.

In a famous passage of the *Philosophical Investigations*, Wittgenstein writes:

> Words are connected with the primitive, natural expressions of sensation and used in their place. A child has hurt himself and he cries; then adults talk to him and teach him exclamations and, later, sentences. They teach the child new pain-behaviour. "So you are saying that the 'word' pain really means crying?" – On the contrary: the verbal expression of pain replaces crying, it does not describe it.[15]

Something similar applies to the relation between the compulsion to repeat and infinite regression: the logical "and so on" *replaces* the compulsion of drives rather than describing it. It is precisely this substitution that permits a new experience of the iteration, a different behaviour with regard to the "one more time." Certainly, at times, even an adult screams in pain; and certainly, even the shrewdest speaker can be dominated by the compulsion to repeat. This does not in the least diminish the importance of the following fact, which is really relevant to an anthropology inspired by naturalism: the life of the human animal distinguishes itself from the life of other animals because of the retroaction of the symbolic plane on the sub-symbolic; because of the replacement of the scream of pain with equivalent propositions, and of the compulsion to repeat with infinite regression.

The hypothesis of a feedback process helps us to explain in a more precise way the peculiar mixture of innovation and restoration that characterises any "and so on" without outcome. In infinite regression, just as in the experience of someone who has fallen prey to the compulsion to repeat, change and the "the conservative nature of living substance" enter into conflict. And yet, in regression, change is not caused by "external forces" but by syntactic recursion, that is to say, by a factor that is inherent to the nature of the specific living being that is *Homo sapiens*. In addition to this, in regression, the very resistance to change assumes a linguistic aspect, presenting itself as a semantic paralysis or as the eternal return of an identical declarative content (the return of the same problem or limit). Therefore, verbal language hosts in itself the whole *oscillation* between development and conservation, achievement of "a state of things which has never yet been attained" and restoration of "an earlier state of things." Unlike other living organisms, in the human animal development is not induced by external causes and conservation is not always prevalently related to the drives: both poles pertain to the way of being of our species; both are rooted in the symbolic plane. What is typical of infinite regression – generated by recursion but, let us repeat this point, by a kind of recursion that retroacts on the compulsion to repeat and remoulds it – is rather the fact that the oscillation between the achievement of "a state of things which has never yet been attained" and the restoration of "an earlier state of things" contracts to the point of collapsing into the immediate *coincidence* of the two antipodes: the *n*th reply of what has already been takes the appearance of novelty, just as novelty entirely resembles what has already been. Thus, the compulsion to repeat, reorganised as it is by syntactic recursion, transforms itself into a constraint that, while remaining imperious – and often equally able to inflict pain – is also of an essentially different kind: it becomes a *compulsion to ambivalence.*[16]

Infinite regression is an exclusively linguistic phenomenon that, however, is able to exhibit the juncture between language and the drives. It is a complex symbolic creation that nevertheless accounts for the way in which the symbolic field (syntactic recursion) retroacts on the sub-symbolic field (compulsion to repeat). We could therefore say that the logical "and so on" is a circumscribed element of human nature and, *at the same time*, the geographical map that represents important features of this nature, that is, of the totality to which it belongs. But there is more to it. Regression, which as such is a result of verbal creativity, that is, of the "infinite use of finite means," does nothing other than emphasise, on the other hand, the permanent interaction between creativity and stereotypy, novelty and repetition. What can an anthropological enquiry gain from the *compulsion to ambivalence* inherent to the interminable "and so on"? In other words, what can it gain from the fact that the achievement of "a state of things which has never yet been attained" is also, at the same time, the restoration of "an earlier state of things" (and vice versa)? The alternative between novelty and repetition usually prepares and substantiates the dichotomy between culture and biology. According to a traditional opinion, which should not be revered, culture would be innovative and biology conservative. Many authors have claimed the exact opposite: culture would stabilise and render consistent the behaviours of the human animal, while biological drives would condemn them to unpredictability. These assertions are perfunctory and untrustworthy. However, it is interesting to observe that, independently of which opinion one privileges, in both cases infinite regression – revealing a logical link between novelty and repetition – constitutes an immediate synthesis of culture and biology. Yet it is insufficient and even misleading to speak of a synthesis. The *compulsion to ambivalence*, which is the true emblem of regression, rather signals (albeit in a paroxysmal and uncanny way) the lack of distinction between culture and biology, that is, the biological character of culture and the cultural character of a biology that is specifically human. I will return to this point below.

a product of the imagination

The compulsion to repeat implies a failure of the imagination, the collapse of "the faculty of representing [...] an object that is not itself present."[17] Fostering the eternal return of the same means, among other things, securing at all costs the presence of the object to be represented, or also, likewise, restricting representation to those objects whose presence is always guaranteed anew. The constant reiteration of the same gesture, opposing the overabundance of environmental stimuli and the change they incite, has an iconoclastic vocation: the only "image" permitted by such a fatal gesture is its actual replication (hence nothing that is really worth being considered as an image). The same that returns is both an event to be described and the description of an event; a figure and that which is figured. Distinguishing itself also in this respect from the compulsion to repeat, infinite regression is, on the other hand, a genuine product of the imagination. The ascending hierarchy of metalanguages, or the inexhaustible succession of new and more comprehensive ideas of Man, is based precisely on the faculty of representing objects that are not present: a metalanguage yet to come; an even broader idea of Man. What is at stake in any interminable backwards march is the immediate intuition of a series whose salient characteristic amounts to remaining always unrealised, defective, and thus only imaginable. When we say "and so on, to infinity," we take for granted the ability to familiarise ourselves with something that, not limiting itself to a provisional absence, will always and at any rate result in being missing; in the end, something that draws its logical and anthropological meaning precisely from a chronic *unpresentability*.

Although it is a product of the imagination, infinite regression is nonetheless not comparable with one or more determinate images. The Kantian notion of *schema* is pertinent to it. Let me explain this analogy in a few words. For Kant, a schema is a "rule" constructed by the imagination which allows us to apply the concepts of the discursive intellect to empirical phenomena. Here is an example: the schema of the triangle (in general), locating itself halfway between a verbal definition of this geometrical figure and the concrete perception of a single triangle (say, isosceles or right-angled), guarantees a juncture between the two. The schema-rule does not represent a particular object but the way in which we represent a class of objects; it is a "monogram of pure *a priori* imagination, through which, and in accordance with which, images themselves become possible."[18] Shortly after, Kant neatly distinguishes between two heterogeneous eventualities. When dealing with "sensible concepts," for instance those of a dog or a lake, in addition to being already in itself a product of the imagination, the schema-rule also functions as a matrix for actual images (of a well-individuated dog or of an unmistakeable lake). Yet things work differently whenever we use "pure intellectual concepts," that is, concepts that are deprived of a perceptual counterpart: for instance, the concepts of substance and causality. In this second case, the schema-rule does not function as a premise of the elaboration of empirical images but constitutes, as such, the ultimate outcome of the work of the imagination. Substance is determined imaginatively as "permanence of the real in time";[19] causality as "the real upon which, whenever posited, something else always follows."[20] But the fact is that permanence and succession cannot in turn be translated into well-inlaid images: they are certainly iconological schemes but of a kind that should exclude the formation of any concrete icon.

Let us return to the relevant point. Infinite regression is undoubtedly a schema, or a "monogram of pure *a priori* imagination." However, it has nothing in common with the concepts that refer to sensible beings (dog, lake, triangle, etc.). Rather, it is one of those schemas that imaginatively determine the "pure intellectual concepts" (substance, causality, etc.). What is precisely the pure intellectual concept that finds its adequate schema in infinite regression? Syntactic recursion; or better: the recursive iteration of a logical circle (overcoming again and again a limit that is nonetheless confirmed by every successive overcoming). Like all other schemas of pure intellectual concepts, the "and so on" is itself a product of the imagination which, at the same time, does not give rise to any real image. Furthermore, like all other schemas of this kind, infinite regression is itself a peculiar way of articulating time (that is, for Kant, "internal sense"). We know that the schema of substance is permanence; that of causality is succession; but, according to Kant, an explicit temporal content equally characterises the schemas of quantity, possibility, necessity, etc. As for infinite regression, that is, the imaginative schema of syntactic recursion, it manifests the interweaving of irreversibility and the eternal return of the same, linear time and cyclical time, or better, their constant juxtaposition. Such interweaving finally coagulates into the figure of the *spiral*, whose new and wider turns propose the initial situation again and again.

On closer inspection, the interminable "and so on" also constitutes the schema of another pure intellectual concept, which is related perhaps to syntactic recursion, but certainly does not coincide with it. The concept in question is nothing other than that of . . . imagination. Here, we are not talking about the concrete operations of this faculty, but of verbal thought's representation of it. The *concept* of imagination must itself have a schema, that is, an imaginative determination. This should not surprise us: after all, it is inevitable that imagination also applies to itself (or better, to the way in which the discursive intellect elaborates its notion), thus producing a "monogram" that corresponds to its own functioning. As a matter of fact, infinite regression, marked as it is by the coexistence of innovation and repetition, is precisely such a "monogram." The concept of imagination is indeed composed of two fundamental elements: *spontaneity*, or the capacity to represent something new that does not have any precedent or prototype; *receptivity*, or the capacity to re-present something that has already been experienced in the past. Spontaneity is the distinctive sign of productive imagination; receptivity that of mimetic or reproductive imagination (as a support of voluntary memory). The "and so on" offers a faithful portrayal of these two attitudes and shows their interaction. On the one hand, it prefigures further and always new steps of an ascending hierarchy that does not prefigure an exhaustive apex (spontaneity). On the other, it tirelessly re-presents an identical initial state of affairs – the very same problem, or limit – (receptivity).

We could say that infinite regression is a product of the imagination that thematises the very prerogatives of the faculty in which it originates. With the proviso that, in regression, spontaneity and receptivity not only operate simultaneously but flow into each other and, most importantly, *paralyse each other*. The representation of the new (spontaneity) has as an inevitable result the re-presentation of what is already known (receptivity); this re-presentation, in turn, foreshadows something that is unprecedented. The prefiguring of a state that was never obtained before (productive imagination) is reversed into the mimesis of an old situation (reproductive imagination), just as the mimesis is turned into a prefiguring: both are thus condemned to a Sisyphus-like task or to a prompt paralysis. In addition to rendering the two components of the imagination almost indiscernible, the circular link between spontaneity and receptivity deprives each of them of their distinctive power: spontaneity, paving the way to a "one more time," does not give rise to any authentic innovation; receptivity, assuming the appearance of an innovative movement, is not able to preserve (or re-present) anything. An identity that – more than ever – is volatile stands as a counterpoint to a difference that is only apparent. Infinite regression schematises the faculty of imagination, but it does so in such a way that it also, and maybe especially, emphasises its eventual *impotence*; an impotence that is caused, let us stress it, by a very strict relationship, or better, a complete overlapping, between its two different components.

At the beginning of this section we observed that the formula "and so on, to infinity" refers to a sequence whose successive episodes, in so far as they are by definition not present (or better, *unpresentable* in their totality), are exclusively the object of the imagination. But the fact that the sequence can only be imagined does not imply in any possible way that it can really be imagined. The ascending hierarchy of metalanguages or of the ideas of Man, starting from a certain point, becomes unimaginable; even its definitive completion is, and forever remains, unimaginable. Infinite regression dilates "the faculty of representing [. . .] an object that is not itself present" but, dilating it, at the same time marks its *limit*. The "and so on" is a genuine product of the imagination, whose decisive property nonetheless amounts to *representing the catastrophe of imaginative representation*. Kant writes that "the imagination succumbs to this advancing towards what is immensely distant, whereby the furthest world is always in relation to one that is even further, and the most remote past always in relation to one which is even more remote."[21] Similarly, imagination succumbs when, wishing to schematically depict the very fact that we speak, it gets caught in the interminable backwards flight of the metalanguages, each of which always presupposes another of a higher degree. Or also when, trying to fix the relation between a singular individual and a species, it transits from a first idea of Man to a second and more comprehensive one, and then to a third, and so on, without a conclusion.

Let me sum up and conclude. Infinite regression is the schema, forged by the imagination, of a pure intellectual concept. More precisely, it is the schema of syntactic recursion (yet the latter is grafted onto a logical circle); or also, in the second place, it is the schema of the reciprocal relation between spontaneity and receptivity (a relation whose strictness transforms it into a paralysing symbiosis). Now, precisely because it is a schema made in such a way, infinite regression enables us to imagine the limit that the imagination is subjected to. It is the iconological construction that emphasises the insufficiency, or defectiveness, of iconic representation. Although it is a "monogram of pure *a priori* imagination," in the final instance regression shows itself to be an *iconoclastic* inclination. Thus it seems to rejoin the compulsion to repeat. However, there remains a macroscopic difference between the two. While the compulsion to repeat is iconoclastic from the beginning since it excludes from itself the work of the imagination, regression becomes iconoclastic to the extent that it brings this work to its *limit*. We could speak of absolute iconoclasm in the case of the repetition of the drives and of a second-degree iconoclasm, mediated by the full unfolding of the imaginative faculty, in that of the unending "and so on." This essential difference between the two forms of iconoclasm substantiates the hypothesis we discussed above: infinite regression does not express or prolong the compulsion to repeat but *replaces* it.

duplicity of aspect

So far we have discussed in detail the hierarchically stratified character of infinite regression and syntactic recursion, leaving mostly aside the kernel of experience that re-emerges with monotonous insistence at each of its stages and the logical (but also pragmatic and emotive) circle in which syntactic recursion is rooted. In order to tentatively denote such a kernel of experience, just as the logical (pragmatic, emotive) circle that characterises it, I have repeatedly made use of two formulas that are both general and provisional. I have spoken of a problem which is deemed to be repeatedly proposed thanks to the very solution we adopt to get rid of it. Furthermore, I have also spoken of a limit that is confirmed each time anew by its overcoming. I have derived some examples from the metaphysical tradition that are able to render these formulas more concretely: Plato's "third man"; the search for an unconditioned condition of empirical phenomena; the mental image we have of our mind; the inevitable metamorphosis of any metalanguage into an object-language. It is now high time to clarify, at least in brief, the anthropological root of those inferences (just as of those actions and passions) for which the conclusion functions as a premise of its premise, that is to say, the naturalistic foundation of the countless instances in which the solution reproduces the initial problem and an overcoming ends up restating the limit that it seemed to have left behind.

The logical or pragmatic circles that, recursively reiterating themselves, give rise to an infinite regression have their common origin in the relation of the human animal to the environment. To be more precise, they have their common origin in the three prerogatives that allow this animal to adapt to a vital context: *hyper-reflexivity*, that is, the biological necessity of representing one's own representations and intervening operationally in one's operations; *transcendence*, that is, the biological necessity of projecting oneself beyond the here and now in order to cling to it, of detaching oneself from one's life in order to continue to live, of always being outside oneself in order to attain a gleam of identity; *duplicity of aspect*, that is, the biological necessity of an artificial or historical-cultural existence, which is, however, extra-biological. From each of these prerogatives we can deduce the other two. To a large extent, we are dealing with synonymous concepts which I certainly do not intend to examine in detail. Firstly, they constitute a consolidated asset of twentieth-century philosophy and anthropology (which only Sperber and Pinker seem to ignore). Secondly, the only thing that interests me here is the relation between the three bio-anthropological prerogatives and the phenomenon of infinite regression. I will therefore speak of hyper-reflexivity, transcendence, and duplicity of aspect only as needed to indicate the way in which they promote the "and so on" without an outcome.

To cut a long story short, I will limit myself to mentioning an observation made by Arnold Gehlen that can also be shared by those who refuse to believe this author's main theses. Our organism, Gehlen says,[22] is subjected to a "flood of stimuli" which, being deprived of an immediate biological finality, do not as such prescribe actions that are advantageous from the perspective of self-preservation. Unlike what happens to the other higher animals, in the case of *Homo sapiens* there is a lack of bi-univocal correspondence between what one perceives and what one does. Rather, disproportion and discontinuity dominate. The overabundance of impressions and stimuli coming from the context is not at all translated into a thorough catalogue of vital tasks. The meaning of the perceptive "flood" remains undetermined, or better, only potential. The permanent gap between stimuli and action induces a certain lack of adherence, or even an actual *distancing*, of the human animal from the states of affairs that surround it. This distancing, together with its subsequent operative uncertainty (and the primordial sentiment of *shame* that it generates), lies at the heart of the three adaptive preconditions to which I referred above.

In so far as its perceptive impressions do not dictate univocal behaviours, in order to survive the human animal needs to control and form them always again by means of a hypertrophic development of reflexive performances. Given that representation cannot found its validity on a pre-existing stimulus, it secures it by becoming itself the object of a second and more powerful representation which is able to consider it and evaluate its perspicacity. Metarepresentation compensates always anew for the discontinuity between environmental stimulus and cognitive response. It retrospectively fills in the void that such discontinuity has inserted into experience. We could say that metarepresentation stands for the stimulus, taking on the orientating function that the latter fulfils in other living species. Obviously, all this is also valid in the case of metaoperations, that is, this practical (working, institutional, etc.) conduct that has its exclusive point of application in other previous and more immediate practical conduct. Reflexive performances, far from being peripheral and exceptional (as happens in animals framed in an ecological niche which functions as a prosthesis for their organism), constitute a primary biological resource. But as long as any reflexive performance has itself to attain a legitimisation it also requires a new and higher reflection. The initial problem, that is, the failed correspondence between the "flood of stimuli" and biologically efficacious behaviours, is proposed again by its supposed solution. The failed correspondence, projecting itself onto the symbolic plane, manifests itself in the ascending hierarchy of metarepresentations and metaoperations. Its presence is even guaranteed by the impossibility of exhausting such a hierarchy – for example, by the impossibility of arresting the march backwards caused by the attempt to elaborate a mental image of one's mind. *Hyper-reflexivity* nourishes the logical and pragmatic circles from which, given certain conditions, infinite regression arises.

The chronic distancing from the environmental context, due to the non-translatability of perceptive stimuli into a detailed stock of advantageous actions, renders the human animal incredibly familiar with *transcendence*. Let us not be misled by this term; what is at stake is only the biological prerogative of an animal that, in addition to living, also needs to make its own life possible. First and foremost, transcendence means that, to the extent that it provides for its own existence, *Homo sapiens* locates itself *beyond* itself as a mere existing being (or *below*: bare existence can in turn result in being transcendent with regard to the set of efforts made to prolong it). The distancing from the environmental context also entails a distancing from oneself as an integral part of that context: the I as spectator overcomes the I as protagonist (or is overcome by it; in any case, it does not coincide with it). Secondly, transcendence means that *Homo sapiens* thematises the *limits* of its field of action. The other zoological species, adhering to their ecological niche (in which there is a proportional relation between perceptions and behaviours), do not sense their limits. On the other hand, the human animal, because of its distancing, senses the limits of the context in which it is situated, and precisely for this reason does not have an ecological niche, that is, an environment in a strict sense, but a historical world. Thematising the limits does not at all imply possessing notions about that which overcomes them, but the opening of a (bio)logical space whose syntactic operator can be referred to as a "beyond of" or "outside of." This operator is completely deprived of any autonomous semantic content and is only useful in the elaborations of efficacious conduct *within* a given context. Transcendence is an adaptive apparatus. But the point is that such an apparatus inevitably generates an exchange of roles between premise and conclusion. The I as spectator, which transcends the I as protagonist, is itself transcended and thus reduced to the role of a protagonist by another I that is even more hidden and registers its movements. Something similar happens in the other form of transcendence. The "environment" of our species is composed of two factors: the environment and its limits, or briefly, $E + L$. However, if the immediate environment is $E + L$, the human animal, which by definition always senses the limits of its environment, also senses the limits of $E + L$; we therefore obtain $(E + L) + L$ as a new and more credible immediate environment. Yet even the latter is defined not only by some concrete characteristics but also by its limits: thus, we will now have to understand $[(E + L) + L] + L$ as environment. And so on, incessantly. The transcending of the vital context is the kernel of experience that infinite regression articulates in the guise of an ascending hierarchy or of a spiral.

The *duplicity of aspect* of the human animal (I draw this phrase from Helmuth Plessner[23] but use it in my own way) is an implicit consequence of hyper-reflexivity and transcendence. As a matter of fact, these bio-anthropological prerequisites already introduce a dense network of splittings and divisions: floods of stimuli that are not biologically aimed at something and the elaboration of advantageous conduct (or, undifferentiated *power* and univocal *acts*); representation and metarepresentation; mere living being and guarantor of one's life; protagonist and spectator; interiority and exteriority; below and beyond; environment and world. The splittings I have just enumerated (and, of course, others could be added to the list) participate in various and variable ways in determining the actual duplicity of aspect: the latter amounts to the fact that man is a *naturally artificial* animal, an organism whose biologically distinctive trait is culture. The distancing from its vital context obliges our species to establish a supplementary relation with it. Now, here is the relevant point (which already emerged in a different way when we talked about transcendence): the human animal is, *at the same time*, a relation and a correlated term, an objective means and a polarity in need of a connection. If we write xRy, that is, "x is in relation to y," we need to specify that in this case "x" coincides with "R," and "R" is not different from "x." A correct way of denoting this might be xXy. The identity, which is, however, also a difference, between "x" as correlated term and "X" as relation illustrates well the identity-difference between biology and culture, nature and artifice, the individual and the social mind. The unity of the two aspects only manifests itself in their gap; and, vice versa, this gap precisely attests to their unity: "x" (the biological organism or individual mind) is equal to "R" (the historical-cultural relation with the operative contexts, or the social mind) precisely in so far as it nonetheless remains distinguished from it. For a materialist anthropologist it is unreasonable to deny the unity of biology and culture but also to misrecognise the gap that exists between them: what really matters is the *inseparability of unity and gap*.

The duplicity of aspect creates an unstoppable circle within which every determination transcends the other and is transcended by it: if it is natural, it then turns out to be really artificial; but if it is artificial, it is then genuinely natural. The historical-cultural actions would have to alleviate the very high degree of contingency, that is, of unidirectional potentiality that characterises the "flood of stimuli." However, inasmuch as they do not derive from a precise environmental signal, these actions have an unforeseeable outcome, and in their turn increase the contingency and potentiality from which they had intended to protect us. Thus, what we need are new historical-cultural actions that, carrying out a metaoperational task, reorganise the relation (R) between the single human animal (x) and the vital context (y). But these second-degree actions themselves expand contingency; and so on. Even in the case of the duplicity of aspect, the problem is reproduced on a wider scale by its solution. Even the circular relation between biology and culture can transform itself into a spiral whose turns are always wider and thus take on the appearance of infinite regression.

I have already observed that a procedure (or rule) is said to be "recursive" if it can be iteratively applied to the result of one of its previous applications. We are now able to specify what the three procedures are, or better, the three bio-anthropological prerogatives that, always applying themselves again to the situation they have generated, give rise to infinite regression: hyper-reflexivity, transcendence, and duplicity of aspect. It is important to add that only syntactic recursion renders these prerogatives species-specific, that is, properly human. Undoubtedly many other animals are capable of reflexive performances, of surviving in an environment that is different from the original, planning some kind of artificial mediation with their operative field. The authentic discriminating factor lies in the tendency to reiterate metarepresentation recursively, the distancing from one's here and now, and the construction of a cultural relation with the context. Without recursion, without the ascending hierarchy of logical levels that it models, we could not speak of infinite regression or, to be precise, of hyper-reflexivity, transcendence, and duplicity of aspect, that is, of the kernel of experience that this regression *reproduces* (receptivity) and *innovates* (spontaneity) always anew. Syntactic recursion is an intra-linguistic property: it concerns the connection between sentences, not between sentences and the world. And yet it is precisely recursion, and not the denotative vocabulary, that moulds the prerogatives thanks to which the human animal adapts to the world. We witness here a peculiar displacement that, on closer inspection, characterises the union of logic and anthropology as a whole: an immediate pragmatic-existential value pertains only to the functions that govern the inner life of verbal language, its relation with itself; in brief, only to what language *is* (syntactic rules and structures), and not to what it *represents*. The organism's impulse of preservation first and foremost avails itself of those traits of human eloquence that are the furthest (and most independent) from somatic and sensory-motor impulses.

Hyper-reflexivity and transcendence resolve themselves into the duplicity of aspect, finding their most adequate expression in the latter. The metarepresentative activity and the capacity of sensing the limits of any environmental context contribute to establishing *Homo sapiens* as a naturally artificial animal. The tendency to operatively intervene in one's operations, as much as the tendency to detach oneself from the immediate here and now, finally coalesce in the identity-difference between biology and culture.

notes

1 See especially *Mondanità. L'idea di "mondo" tra esperienza sensibile e sfera pubblica* (Rome: Manifestolibri, 1994); *Parole con parole. Poteri e limiti del linguaggio* (Rome: Donzelli, 1995); *Il ricordo del presente. Saggio sul tempo storico* (Turin: Bollati Boringhieri, 1999).

2 See *Parmenides* 130a–134c.

3 *Metaphysics* 990b 8–15.

4 I. Kant, *Critique of Pure Reason* (Basingstoke: Palgrave, 2003) 386.

5 Ibid. 391.

6 See R. Rucker, *Infinity and the Mind: The Science and Philosophy of the Infinite* (Boston: Birkhauser, 1982) 45–46.

7 S. Freud, *Beyond the Pleasure Principle* in *The Standard Edition of the Complete Psychological Works of Sigmund Freud* (London: Hogarth, 1955) XVIII: 36.

8 Ibid.

9 Ibid. 38.

10 N. Chomsky, *Syntactic Structures* ('s-Gravenhage: Mouton, 1957); idem, *Aspects of the Theory of*

Syntax (Cambridge, MA: MIT P, 1965); idem, *Reflections on Language* (New York: Pantheon, 1975).

11 On this issue see H. Putnam, "Ricorsività" in *Enciclopedia Einaudi XII* (Turin: Einaudi, 1981) 33–61.

12 N. Chomsky, *Language and Problems of Knowledge: The Managua Lectures* (Cambridge, MA: MIT P, 1988).

13 Freud, *Beyond the Pleasure Principle* 41.

14 Ibid. 38.

15 L. Wittgenstein, *Philosophical Investigations*, 244 (Oxford: Blackwell, 1953) 95.

16 See M. Mazzeo, *Contraddizione e malinconia. Studio sull'ambivalenza dell'animale umano* (Macerata: Quodlibet, 2008).

17 Kant, *Critique of Pure Reason* 165.

18 Ibid. 183.

19 Ibid. 184.

20 Ibid.

21 I. Kant, *Critique of Practical Reason* (Cambridge: Cambridge UP, 1997).

22 See A. Gehlen, *Der Mensch. Seine Natur und seine Stellung in der Welt* (Berlin: Junker, 1940). See also M. Mazzeo, *Tatto e linguaggio. Il corpo delle parole* (Rome: Riuniti, 2003).

23 H. Plessner, *Die Stufen des Organischen und der Mensch. Einleitung in die philosophische Anthopologie* (Berlin: de Gruyter, 1928).

roberto esposito

translated by lorenzo chiesa

POLITICS AND HUMAN NATURE

I

The *Letter on Humanism*, which Heidegger published in 1946 in the midst of a historical and biographical defeat, seems to close the age-old story of humanism. Despite all attempts at restoring it in spiritualistic, Marxist and existentialist guises – Sartre's essay *Existentialism is a Humanism* was released in the same year – the great humanist tradition could not resist the double trauma of Auschwitz and Hiroshima in which the very idea of humanity had been swallowed up by its opposite. Leaving aside all the circumstantial and even instrumental elements that determined the drafting of the *Letter*, the necessity of such an epistemological break lies at the centre of Heidegger's text: a human culture that has not been able to avoid, or even favoured, the slaughtering of fifty million people around the middle of the twentieth century cannot expect to survive. After a catastrophe of these dimensions, the idea that it is possible to reinstate the old humanist myth of man as master of his destiny is doomed to fail for at least two sets of reasons: firstly, it is impossible to go back in history to a period that is irreparably terminated. Secondly, the smoking rubble that at the end of the war occupies the field in both a symbolic and material sense originates precisely from this period.

And yet, although Heidegger's *Letter* marks a point of no return with regard to everything that the idea of *humanitas* meant for at least five centuries, we cannot say that he entrusts it to a language which is truly innovative. This is due not only to the author's temporary compromise of principles with the powers of the anti-human but also to a more intrinsic reason, one that is inherent in his very definition of "humanity." While we are not able to investigate this question in detail here, we could say that it is precisely the modality of Heidegger's separation from the humanist tradition – which he unifies following a thread that links the Greek *paideia* to the *romanitas* up to the modern *studia humanitatis* – that keeps him within its semantic scope. After all, he himself stated that it is necessary to think against "humanism [. . .] because it does not set the *humanitas* of man high enough."[1] What did Heidegger mean with these words? In what sense would humanism have betrayed itself, placing man at a level that is inferior to that of his innermost meaning? The philosopher's answer is well known: humanism has not been able to emancipate itself from the vocabulary of metaphysics because it "thinks of man on the basis of *animalitas* and does not think in the direction of his *humanitas*."[2] Humanism understands man as an animal species, which is certainly distinctive, since it is endowed with the gift of reason, yet not fundamentally different from others. Independently of the name that has been assigned to him, he has always been conceived as an *animal rationale*: "In principle," Heidegger continues, "we are still thinking of *homo animalis* – even when *anima* [soul] is posited as *animus sive mens* [spirit or mind], and this in turn is later posited as subject, person, or spirit [*Geist*]."[3] According to Heidegger, this was the fatal error that has led humanism, first, to contradict and, then, turn against itself; not the search for a human essence that precedes man's existence, as Sartre would still have had it, but the failure that involves not distinguishing him from a living matter which is not specifically human, or even deriving him from it. From this point of view, the bestialisation of man, tested in the Nazi extermination camps, would find its roots in the categorical confusion between man and animal with which the humanist concept of *humanitas* has been characterised since the beginning.

According to Heidegger, this indistinction, or union, between animality and rationality deprives the man of humanism of his privileged relation with the sphere of being, which is, on the other hand, only recognisable in the element that radically separates humankind from any other living being – that is to say, language. To state that it is language that defines man as such means, for the German philosopher, to define him precisely on the basis of his insurmountable opposition to animal dumbness. But precisely for this reason, it also means to push the phenomenon of simply being alive [*semplice vita*] that links all living beings in the same biological dimension outside of the field of the definition of man. As soon as this path is taken – that of the subtraction of man, or at least of his ultimate truth, from the dimension of life – Heidegger can put forward the hypothesis that "the essence of divinity is closer to us than what is foreign in other living creatures."[4] In spite of an early interest in "factical life" manifested already in the Freiburg courses of the 1920s, and of a certain curiosity with regard to medicine and psychiatry witnessed by the Zollikon seminars, we can say that the abyssal divergence Heidegger establishes between man and animal is what maintains his philosophy at a great distance not only from what the Greeks called *zoé*, mere life, but also from the whole horizon of *bíos*. The fact that the animal is then defined as "poor in world" – unlike man who is "world-forming" – is another way to mark the irreducible gap that, in *Dasein*, separates the sphere of being-in-the-world from that of biological life. On the other hand, what makes man a being worthy of such a name is not life, which is also present in all inferior organisms, but rather death, which, unlike man, the latter undergo unwittingly, without ever being able to surmise it. Man is essentially mortal, rather than living; he is a being-for-death. This is the element that Heidegger puts at the centre of his ontology in opposition to the humanist tradition, but also, at the same time, to any positive knowledge of life. His thesis is that in order to grasp human reality more deeply than humanism has ever done it is necessary to think of it outside of the common horizon of what merely lives. For Heidegger, man's truth lies beyond, or before, his mere life – this is also demonstrated by the philosopher's fundamental lack of interest in the sphere of the body. For this, although he claims to be aiming at focusing his reflection "not only [on] man but also [on] the 'nature' of man," he excludes from such nature any biological reference; rather, he identifies precisely in this exclusion the way that leads to the most primordial dimension "in which the essence of man, determined by Being itself, is at home."[5]

2

Can we say that Heidegger has really been able to get rid of the tradition which he intends to criticise? That he speaks a language entirely different from that of classical humanism? It is difficult to answer affirmatively to these questions. Certainly, the decentring of the human being [*ente umano*] from the dimension of being [*essere*] around which Heidegger's whole thought revolves is all too evident; just as his deconstruction of the subjectivist – and objectivist – metaphysics which entrusts man with the ownership of his own self and of all that surrounds him is adamant. *Dasein* is not the man of humanism, at least because he is constitutively split by a difference with regard to himself that deprives him of any solidity and stable consistency. Yet all this occurs in a manner that, instead of contesting man's absolute uniqueness and deviation from other living beings, confirms and develops them in a way that is not so distant from traditional anthropodicy. Having been abandoned by God to his earthly destiny, Heidegger's existent [*l'esistente*] inherits from him not only his absoluteness but also his primacy over any other species. He is not separated from them by a different nature, but precisely by his structural lack: the essence of man lies more in his non-involvement in the natural field than in his belonging to a specific nature. It is here that the conceptual dichotomy between natural sciences and the knowledge of man is rooted and intensified, in spite of Heidegger's intention to overcome it through a different theoretical vocabulary. The reality of man – existence – is not recognisable by means of the natural sciences because man does not have, properly speaking, a nature; or because his nature is, as a matter of fact, essentially unnatural. Rather, he has a "condition," as Hannah Arendt will say from a different point of view in *The Human Condition*, sharing the anti-biologistic bias of her teacher: life is the biological presupposition from which human existence must part in order to assume the – anthropological, political, philosophical – meaning that belongs to it. But it is precisely this anti-biologistic bias that retains Heidegger's *humanitas* within the framework of the humanist tradition, especially that of Pico della Mirandola, which, at the same time, he would like to supersede. Was it not precisely Pico who located the *dignitas humana* in a condition of eccentricity with regard to any natural given, and also identified precisely in it the difference, as well as the superiority, of man over all other animals? While the latter are granted by God a determinate nature – and are inextricably linked to a natural environment, as twentieth-century philosophical anthropology has explained – man, and man alone, does not have a set place in the world, and precisely for this can choose it according to his preference. In this way, free to degenerate into inferior beings as well as rise up to the level of divine beings, he is able to transform himself continuously: "Fingit, fabricat et transformat se ipsum" according to his own will.[6] There is no ontological constraint, fixed character, or natural invariant that binds him to a specific natural modality. He is not nothing, since he can become anything, create himself again and again according to his own liking. Properly speaking, he is not even a being, but a becoming in perpetual change.

When, after four centuries – giving the impression of opening a whole new scenario – Sartre writes that "man is free and there is no human nature in which I can place my trust,"[7] he only proposes again the original idea with which the humanist tradition set itself in motion. Obviously, he does so from a point of view that privileges the level of existence over that of essence, yet bringing existence back to the same anti-natural character through which the humanist tradition thought essence. It is sufficient to replace the term "essence" with that of "nature" in order to grasp the actual convergence of two positions that are different only in principle. To assert that "man is nothing other than what he makes of himself"[8] – as Sartre argues – means locating him in a radically historical dimension, depriving him of any natural features which can be presupposed to belong to him. True, man is always finite, but precisely for this he is also infinitely creating himself: he is the subject of his own substance in a form that continuously dissolves substance into his own subjectivity. The introduction of Marxist motifs into his earlier phenomenological background does not greatly change the basis of Sartre's argument: nature is considered to be the material instrument of the human subject's historical self-production, rather than his biological component. When Sartre insists on the fact that, according to existentialism, man is always outside himself, projected towards a continuous transcendence of his own natural condition, he must be taken literally: although man is placed in a series of material conditions that precede him, he experiences his authentic humanity precisely at the point at which he separates himself from them in order to project himself according to his existential decision. His nature matters only in so far as it is overcome. Subjected to a complete historicisation, the dimension of existence ends up being placed at a radical distance from that of life. In other words, life acquires a human character only in so far as it is subtracted from its biological meaning. While with respect to philosophical terminology Heidegger neatly distances himself from this semantics, in the sense that he deconstructs every one of its concepts – from subject to substance, nature to history – from the point of view of the relation between *humanitas* and *bíos* he remains fundamentally within its boundaries. The very temporal determination of *Dasein*, which nonetheless de-structures any full image of subjectivity, goes in the same anti-naturalistic direction. Rather than being truly abandoned, the humanist discourse appears to be integrated – subsumed and sublated – dialectically into a further configuration. None of its starting presuppositions – be it the refusal of the biological notion of human nature, the absolute opposition of man to other living species, or the underestimation of the body as a primary dimension of existence – is really put into question. Beyond the critique of humanism the ancient profile of *man* as essentially *humanus* delineates itself again.

3

But there is another reason why we cannot say that Heidegger inaugurates a new post-humanist language. In all truth, this language did not have to wait for his *Letter*. First Darwin and then Nietzsche had already used it, if not invented it – in manners which cannot be made to overlap and which actually led the latter to bitterly criticise the former, although he had only a second-hand knowledge of his work and basically misunderstood it. What Nietzsche criticised – wrongly identifying it with the positivist doctrine of progress – was not Darwin's original idea but its simplistic interpretation provided by Spencer. We should rather say that the most significant meaning of the Darwinian perspective lies precisely in the deconstruction of the teleological and essentialist progressivism adopted and elaborated by the humanist tradition. While we are not able to dwell on the overall structure of Darwin's theory of evolution here, there are basically two points, clearly interlinked, in which one senses more distinctively its distancing from the humanist vocabulary: on the one hand, the replacement of the search for the essence – or even the condition – of man with the search for a nature defined on the basis of a series of biological invariants; on the other, the positioning of this nature within the general chain of living species, in spite of the peculiarity of its own characteristics. This does not mean that Darwin reduces human behaviour to a mere response to its organic components, or that he opposes history to nature as something stable and unchanging. On the contrary, he joins them together according to a concept of natural history that entails the modification of human nature on the basis of a series of exceptions to the norm which cannot be determined in advance, but are produced spontaneously by chance. It is precisely on these variations that the mechanism of natural selection is exercised. This constitutes the third and most prominent element of divergence from the humanist tradition: human nature is not a whole that progresses towards what is better, but the always modifiable result of an inexhaustible conflict between different biological typologies which compete with each other in order to affirm themselves.

As is well known, Nietzsche assumes this very principle and brings it to its extreme consequences. Again, we are not able here to follow his argument in all its infinite folds, nuances, and contradictions. But at its centre certainly lies his awareness of European humanism's backwardness and inadequacy with respect to what appears to him to be the most important requirement of his, and our, age (an inadequacy he defines as nihilism precisely in so far as it always intends to strengthen its own exhausted values). That is to say, not only the scientific, philosophical and political centrality of the *bíos* as a whole and in its internal thresholds but also the struggle that is being fought concerning the ways and the outcome of its transformation.

What Nietzsche perceives more lucidly than any other thinker of his time is the fact that behind, and inside, the classical question of *humanitas* a conflict erupted whose stake is the very definition of man. But also of what he might become at the moment when the question of his modified production is posed. From this point of view, we should say that Nietzsche resumes Pico's old myth of human plasticity – man's production of his own essence. Yet, at the same time, he translates it into biological terms to the extent that he assumes the body of man, not his soul or social condition, as the object of such transformation – or better, man as a bio-determined whole in which soul, condition, and body form a single living organism.

Obviously, all this is not devoid of disquieting risks. As soon as this anthropo-technical – or biopolitical, as I prefer to call it – vector of artificial intervention into the characteristics of human nature enters into contact, or synergy, with the other Darwinian presupposition about the contiguity with the animal world, itself deployed in social or even ethnic-racial terms, the consequences can be devastating. When Nietzsche asks himself "Why could we not succeed in doing with humanity what the Chinese have learnt to do with a tree – making it bear roses on one side and pears on the other?,"[9] he is not only theorising the passage from Darwin's natural selection to a project of artificial selection but also prefiguring an anthropological setting in which humankind is divided into non-equivalent categories of selected and selectors. What's more, the comparison with the plant and the animal, also present in other texts, goes in the direction of an ideological drift that aims at assimilating a part of mankind to living species of an inferior kind, and puts it at the service of another part which is, on the other hand, meant for superior performances. Even in this case, Nietzsche appropriates Pico's motif concerning the oscillation of the human being between degeneration towards the animal and regeneration towards a divine dimension. Yet, for him, the different – animal and divine – conditions become internal anthropic [*antropiche*] typologies of the human kind, rather than constituting the polarities between which it oscillates. From this position to the animalisation of a certain kind of man – and the divinisation of another – brought to the most radical and pernicious thanatological consequences in the following decades, is not an impossible or unforeseeable step. Clearly, this does not mean reducing Nietzsche to an outcome for which his thought is in no way responsible – as precisely those who arbitrarily claimed his legacy attempted to do. But terms such as "domestication" (*Zähmung*) and "breeding" (*Züchtung*), referring to groups of men selected to this end, open nonetheless an uncontrolled breach in the old notion of *humanitas* within which any option becomes possible. This action of "domestication" of man with regard to his primordial animal tendencies, which a long and glorious tradition, from Erasmus to Goethe, thought in terms of education and spiritual formation, is now reinterpreted in an anthropotechnical and zoo-technical sense. On the other hand, as we have already noted, what happened in Nazi Germany is not completely external to an obscure background that has been present in the humanist and anthropocentric *ratio* since its beginning. As witnessed by the extraordinary development of German anthropology of that time – which went together with that, parallel and interlinked, of zoology – the Nazis never discarded the category of *humanitas*. And this to the point that a number of handbooks of racial hygiene referred precisely to humanity in their titles: were they not trying to improve humankind by immunising it against its contaminating refuse? As a matter of fact, rather than directly bestialising man, Nazism widened the definition of *anthropos* up to including into it even animals of inferior species. The one who was subjected to extreme violence was not simply comparable to animals. He was an animal-man: an animal with a human face or a man inhabited by an animal.

4

But is this the only – destructive and self-destructive – face of post-humanism? Is it necessary for it to turn into a form of patent anti-humanism? Or does the end of humanism also open up another horizon of meaning in which the classical figure of *humanitas* lends itself to new interpretative possibilities? In any case, we should start from the refusal of any nostalgic – or restorative – approach, like the one adopted by Lukács at the end of the Second World War against the supposed destroyers of reason, and implicitly re-proposed today by Habermas – undoubtedly in a more sophisticated manner. As the German philosopher Peter Sloterdijk has argued, precisely in a dispute with Habermas, it is not only the humanism of the early modern age but also its political re-translation into the national cultures of the nineteenth century that need to be considered as a fundamentally exhausted phenomenon.[10] Contemporary societies no longer produce their political syntheses according to the model provided by the literary society. On the other hand, after the end of the Hitler Youth, it would not have been possible to imagine the reanimation of a novel Goethean youth. Heidegger himself did not address his *Letter* to the German people, as Fichte did in his time, but to a foreign correspondent – in this particular case, the Frenchman Jean Beaufret. We could say that, today, if we still wish to use it, the term *humanitas* can no longer be articulated from the perspective of a national tradition, but, in a different and wider sense which I would like to define, at the same time, as singular and global, referred to every man and the world in its entirety. From this point of view, despite its contradictions and the instrumental use that has been made of it, Nietzsche's thought marks a threshold of awareness behind which we can no longer withdraw. Unlike what has been proposed, or presupposed, by all forms of – lay and religious – essentialism, historicism and personalism formulated before and after him, the humanity of man can no longer be thought outside of the concept, or better the natural reality, of the *bíos*. Singular and collective life, in its requirements of conservation and development, is today the only criterion of universal legitimisation that gives sense to the political, social, and cultural practices of our world. This means that the notion of human nature – which is always more at the centre of scientific as well as philosophical interest – must be thought not in opposition to that of history but in relation to it.

In this sense, we need to resume – evidently, with all the necessary modifications – Darwin's original intuition according to which not only do invariance and mutation not exclude one another but are reciprocal. More precisely, in the human being, it is precisely what is innate that opens up a range of acquisitive possibilities which, in turn, reflect themselves retroactively into genetic programming: we could say that man is programmed to change his programming continuously. The philosopher and historian of medicine Georges Canguilhem was able to claim that, in the case of a human being, health coincides with the organism's ability to continuously change its own norms, rather than with a condition of static normality. On the contrary, illness is nothing but the atrophy, or the weakening, of such a power of innovation.[11] In the end, this is a new way of understanding Pico's motif concerning the infinite variety of human nature, although it is now referred to the body. But such variety needs to be thought as the specific product of the laws of biology, rather than as a miraculous subtraction from them. If this is true, metaphysical dualism of Cartesian origin and bio-sociological reductionism amount to the two opposite and complementary sides of the very same error, which results, in one case, from absorbing the invariant within contingency and, in the other, dissolving contingency in the invariant. While, on the other hand, the proper place of philosophical, but also scientific, enquiry on human nature lies exactly at the point of juncture, or in the margin of indistinction, between natural regularities and socio-cultural variations.[12] It goes without saying that this renders obsolete – and therefore strongly conservative – the opposition between sciences of the spirit and natural sciences, but also the opposition between the empirical and the transcendental: just as the activities of the mind and of language are connected with the organic structures in which they are rooted, so these are in turn modified by the linguistic and mental performances they produce. If, in short, the movement of the hand depends on a command of the brain, the functioning of the brain is itself dependent on the operations of the hand. In this sense, it is evident that the entire course of history is the free and infinitely variable outcome of a bio-natural necessity.

5

But if this is true, the opposite also holds: just as nature powerfully influences history, so history acts retrospectively on nature in an equally significant measure. Here we come to the most complex and problematic aspect of this issue – to what, having examined the "Darwinian side" of the question, we could define as its "Nietzschean side." This is what we previously called the anthropo-technical, or anthropo-poietic, vector, which is increasingly active in today's world. We could suggest that in comparison to Nietzsche's time, and also to that of Nazism's insane biocracy, the possibility of the artificial modification of human nature has been enormously enhanced. I am not thinking exclusively of techno-biology but, more generally, of the drastic decrease in mediations between, on the one hand, the fields of politics, law and economy, and, on the other, the dimension of *bíos*.

When one – rightly – speaks of biopolitics, one precisely alludes to the fact that it is no longer just the faculties that characterise man from the point of view of history which are massively invested by socio-political processes but also those that define him from the point of view of his nature. It is only in this way that we are able to account for phenomena which would otherwise remain inexplicable, such as the new centrality assumed by the ethnic element within current political conflicts and the immediately productive use – in the world of work – of the eminently natural faculty of linguistic communication. In spite of the profound differences between these two phenomena, what is in any case determined is the inscription of practices of political power and economical production within the sphere of *bíos* – or, from a different angle, the overbearing entrance of biological life into socio-political dynamics. When we add to these processes the expansion of bio-technologies as such, allowed by the overwhelming development of genetic and cognitive engineering, the resulting picture may appear very disquieting. By now, we are in a situation which is far more exposed than that described by Foucault's classical analyses of the disciplinary control of bodies exercised by given political regimes, one in which the very subject of biopower tends to broaden and generalise itself into planetary apparatuses that regulate life according to entirely technicised [*tecnicizzate*] procedures.

And yet, in spite of the risks inherent to these transformations, for various reasons, they are not to be understood only with regard to their terrifying effect – such as the posthumous triumph of the inhuman predicted by the "bio-catastrophists."[13] First of all, because technology [*tecnica*] is not necessarily opposed to nature; in fact, in some way, it derives from it, in the sense that human nature possesses a primordial technicity [*tecnicità*], as witnessed by the technical character of any movement of our body and sound of our voice. In addition to this, technology [*tecnica*], all our performances, from the simplest to the most sophisticated, influence our nature retroactively. Thus, from this point of view, measured on the phylogenetic level, every technology [*tecnologia*] is in principle bio-technological. Without doubt, and precisely for this reason, technical skill is not only a production of artefacts but also a transformation of those who produce them – an alteration of man himself beyond that of matter and the environment. This is the most delicate point of the issue, starting from which the whole discourse on humanism could acquire a new and unexpected value. First, humanism is not all the same, not even in its classical form. In opposition to the anthropocentric essentialism of those who considered man as an absolute and inimitable model, authors such as Bruno and Spinoza provided us with a meaning for the absolute multiplicity of forms which human nature can assume. Diversity, alterity, hybridisation are not necessarily a limit and a danger from which we have to defend ourselves in the name of a self-centred purity of the individual and the species – following an immunitarian semantics that has led to the most brutal forms of homicidal eugenics. If these are returned from the blocked and exclusive logic of *immunitas* to that – open and inclusive – of *communitas*,[14] they are also a resource and an opportunity. This is valid both in relation to other living species, such as, first and foremost, animals, and in relation to non-organic forms and matter. The two things should not be separated: that is, the intra-specific relation between man and animal, dating back to the age of domestication, constituted the first anthropo-technical, i.e., self-transformative, segment of the entire process of hominisation [*ominazione*]. For this, contrary to what is presupposed in the bestialisation of the enemy carried out by all – ancient and recent – racisms, engaged as they are in the symbolic fabrication of the man-animal, the relation to the zoosphere, that is, the animal world, has always amounted to a decisive advancement of human culture.[15] In opposition to what Heidegger thought, the animal is not the ancestral past, the stone guest, the mute enigma, but the future of man: it is a place, and a threshold, from which man can only be stimulated in view of a more complex and open elaboration of his *humanitas*. Something along these lines can also be said about technology – to the extent that it does not blatantly contradict the biological possibilities contained in our naturally altered nature, but leads them to their highest development. Evidently, like all supreme opportunities, this is situated on the reverse side of a considerable risk. The history and the destiny of what comes after humanism are written, for now, in letters that are not all decipherable, between the two implications of this argument.

notes

1 M. Heidegger, *Basic Writings* (London: Routledge, 1978) 210.

2 Ibid. 204.

3 Ibid. 203.

4 Ibid. 206.

5 Ibid. 225.

6 Pico della Mirandola, *Discorso sulla dignità dell'uomo* (Brescia: La Scuola, 1987) 8–9.

7 J.-P. Sartre, *Existentialism is a Humanism* (New Haven: Yale UP, 2007) 36.

8 Ibid. 22.

9 F. Nietzsche, *Frammenti postumi 1881–1882* (Milan: Adelphi, 1964) 10.

10 See P. Sloterdijk, "Rules for the Human Zoo: A Response to the *Letter on Humanism*," *Environment and Planning D: Society and Space* 27.1 (2009): 12–28.

11 See G. Canguilhem, *The Normal and the Pathological* (Cambridge, MA: Zone, 1991).

12 See Various authors, *La natura umana* (Rome: Derive Approdi, 2004).

13 See D. Lecourt, *Humain posthumain* (Paris: PUF, 2003).

14 See R. Esposito, *Communitas: The Origin and Destiny of Community* (Palo Alto: Stanford UP, 2009).

15 See R. Marchesini, *Post-Human. Verso nuovi modelli di esistenza* (Turin: Bollati Boringhieri, 2002).

alvise sforza tarabochia

AFFIRMATIVE BIOPOLITICS AND HUMAN NATURE IN FRANCO BASAGLIA'S THOUGHT

I introduction

In this article, I contend that Franco Basaglia's work of reform in psychiatry can be read as a proposal for an affirmative biopolitical psychiatry. My claim is that Basaglia's reform is to be regarded not only as a successful dismantling of the old disciplinary psychiatry, which was based on a purely organic treatment of mental illness and confined within the closed space of the asylum, but also as an *ante litteram* attempt at preventing what Pierangelo Di Vittorio refers to as a "biopolitical psychiatry": by the latter he means a psychiatry that makes extensive use of psychopharmacology, psychiatrises all forms of psychological suffering, reduces the human psyche to its organic correlative – for instance, by focusing exclusively on neuroscientific research. Di Vittorio's definition is notably limited, for he seems to overlook Foucault's conceptual oscillations between affirmative biopolitics and negative biopower. As Roberto Esposito has argued, while affirmative biopolitics enhances, promotes and *fosters* life, biopower amounts to its negative drift, its intrinsic risk of becoming a *thanatopolitics* which ultimately *disallows* life through the very means used to foster it. Di Vittorio's biopolitical psychiatry does not bear any affirmative connotation, and his radically negative perspective should rather be considered as the exertion of biopower in psychiatry, as a negative involution of an affirmative biopolitical psychiatry, which is largely yet to be theorised.

In light of these considerations, I propose to read Basaglia's entire work of reform, which can be summarised by his famous motto "bracketing mental illness"[1] – that is, disregarding the patient's diagnosis – as an affirmative biopolitical psychiatry. This reading discloses for the first time a crucial aspect of his thought. Colucci and Di Vittorio have already pointed out Basaglia's preoccupation with the fact that closing the asylums might ease the conversion of disciplinary psychiatry into a generalised condition, in which the whole population would be "psychiatrised" and "medicalised"[2] – the crucial characteristic of biopolitical psychiatry in Di Vittorio's terms.

I would suggest that biopolitical psychiatry cannot be understood only in negative terms, as Di Vittorio does, and that Basaglia did not propose to oppose its emergence; rather, he actively fought against disciplinary psychiatry. It seems to me that Basaglia proposed to implement biopolitical psychiatry in an affirmative way. Such a proposal would not prevent psychiatry from being one of the *dispositifs* of biopolitics, that is, one of the possible means through which political power reaches the lives of people: this prevention might be impossible or even undesirable. On the other hand, it would preclude psychiatry from becoming a *dispositif* of *biopower*, one that would ultimately bring life under such strict control that it would degenerate into a thanatopolitical implosion.[3]

I will summarise Basaglia's affirmative biopolitical proposal by means of the expression "clinic of lack."[4] Basaglia's bracketing of mental illness entails that all a priori assumptions concerning a human being, such as reducing his essence to his biological nature, should be disregarded. In doing this, Basaglia is indeed implying that all individual subjects lack the possibility of defining their own nature meta-historically, outside of a given social context, and without a relation to the other. This consideration is in itself sufficient to problematise Colucci's reading of Basaglia's proposal for an alternative psychiatry, which, according to Colucci, should be a psychiatry that aims at returning to the subject, and reconstructing a *polis*.[5] If we make good Basaglia's suggestion that the subject is *constitutionally lacking*, the return to the subject can only amount to the acceptance of this structural lack of subjectivity. Consequently, the *polis* to be reconstructed would not only amount to a community of constitutionally lacking subjects but also and especially to a community of subjects who, to a certain extent, *renounce* their subjectivity. I will also argue that Basaglia's acceptation of community can further be understood through Esposito's notion of *communitas*. In this rests the affirmative biopolitical aspect of his proposal: Basaglia is urging the formation of a community that, as far as possible, forsakes its paradigms of *immunisation*, of defence against the loss of individuality that is inherent in the very fact of belonging to a community.

Before embarking on the analysis of biopolitics and psychiatry, I deem it necessary to introduce briefly the life, works, and achievements of Franco Basaglia.

2 franco basaglia

Basaglia (1924–80) began his career as a psychiatric practitioner in the University clinic of Padua, where he worked until 1961.[6] This clinic had a radically organicist approach to psychiatry, whereas Basaglia, under the influence of philosophers such as Husserl, Heidegger, and Sartre, and psychiatrists such as Jaspers, Binswanger and Minkowski, was at the time attempting to implement the tenets of phenomenological psychiatry and *Daseinsanalyse*. Unable to fulfil this task in Padua, he was appointed director of Gorizia's asylum in 1961, where he developed the idea that reforming psychiatry demanded nothing less than the "destruction of the psychiatric hospital."[7] After carrying out several transformations in this direction – such as opening the wards of the asylum, abandoning the white coat, etc. – Basaglia met the resistance of Gorizia's council administration and, after a short stay in Colorno's asylum, moved to Trieste (1971), where he finalised his reform and campaigned for the approval of Law 180, which was ratified in 1978. Law 180 thoroughly reformed Italian psychiatry, which was still regulated by a law dating back to 1904. It did so in three major ways. First, it decreed that involuntary hospitalisation could be enforced for a maximum of seven days, and that such internment would be renewable only after adequate medical examinations. Before Law 180, involuntary hospitalisation was generally regarded as a life sentence, because, once enforced, its duration was at the discretion of the psychiatrist. Second, Law 180 banned the creation of new psychiatric hospitals, and decreed that existing asylums should close. Finally, it entrusted territorial centres (*Centri di Salute Mentale*) and small short-stay hospital wards (*Servizio Psichiatrico di Diagnosi e Cura*) with the provision of psychiatric treatment. As Colucci and Di Vittorio have it, Law 180 was a Copernican revolution, which granted Italy a pioneering role in psychiatric treatment worldwide.[8]

Basaglia himself remarks in an interview that Law 180 was a milestone in psychiatric assistance in so far as it recognised the basic civil rights of the mentally ill.[9] The new legal measures broke the logic that had governed institutional psychiatry and the asylum since the time when, in seventeenth-century Europe, the first madmen were institutionalised along with criminals, the poor and indigents – the era that Foucault refers to as the "great confinement,"[10] which "banished madness into the dull, uniform world of exclusion."[11] Nevertheless, Basaglia also notes that "this Law has somehow raped the very practitioners of alternative psychiatry,"[12] because, as Donnelly puts it, it provided "very limited guidance about how positively to reconstruct 'help' to psychologically suffering" people.[13] Once the reform was achieved, the anti-institutional psychiatrist no longer had any positive and concrete indication of how to manage the new – de-institutionalised – system he was nonetheless still in charge of. Basaglia advanced his view that this lack of method, technique, guidelines, and theories should be regarded as an "ideological void."[14] Yet for this very reason it was also a "happy moment,"[15] during which psychiatrists could be independent from the impositions of any implicit ideology, such as those dictated by organicist or social psychiatry.[16] In Basaglia's words:

> unarmed as we are, devoid of instruments [...] we are forced to establish a relationship with anxiety and suffering without objectifying them automatically in the schemes of "illness" and without having an established interpretative code.[17]

In other words, Law 180 was the ultimate attempt at "bracketing mental illness,"[18] which did not mean denying the possibility that madness could have organic roots, but acting *as if* it did not; omitting "any nosographic definition"; disregarding "illness as an actual fact"[19] in the daily practice of psychiatry. The sick person, no longer regarded as an object to be studied and treated, is approached as an independent subject. His subjectivity "can emerge only in a relationship that [...] manages not to enclose the abnormal experience in a further objectification."[20] That is to say, it can emerge only in a relationship that is not regulated by a priori assumptions.

3 biopolitics and medicalisation

Despite Basaglia's efforts to put into action the "bracketing" of mental illness through Law 180, which banned the asylum and limited the coercive nature of involuntary hospitalisation, "merely changing the law is no magic formula for altering the practice of psychiatrists."[21] More specifically, according to Di Vittorio, the implementation of Law 180 did not prevent the rise of what he calls "biopolitical psychiatry."

3.1 biopolitics, biopower, discipline

Di Vittorio claims that, while the Italian psychiatric reform effectively counteracted the disciplinary aspect of psychiatry, it failed to address what he defines as the "emergence of biopolitical psychiatry."[22] His distinction between the disciplinary aspect of psychiatry and its biopolitical development rests on the distinction between discipline and biopolitics, which Foucault draws – not without some inconsistencies – in his 1975–76 course *Society Must be Defended*. In Foucault's words, biopolitics is "a new technology of power" that does not rule out disciplinary power, but rather "integrates it, modifies it [...] and uses it [...] by embedding itself in existing disciplinary techniques."[23] On the one hand, discipline "is addressed to bodies,"[24] in that it aims at controlling *individuals* outside of the groups they may constitute. On the other, biopower[25] is exerted on the "global mass" of individuals, in so far as they can all be reduced to their shared organic and mental processes.[26] If the disciplinary "seizure of power over the body is an individualising move," the power that biopolitics exerts "is not individualising but [...] massifying."[27] That is to say, it seizes power over man considered as a species, reduced to his biological nature.

The boundaries between discipline and biopolitics are blurry, and it seems to me that Foucault intended them to be so: biopolitics evolves from disciplinary techniques, which it overcomes by, at the same time, remaining embedded in them. Arguably, Foucault seemed to regard the chronological evolution of power in the West as a four-stage process: sovereign power, discipline inside the institutions (e.g., asylums), disciplinary society, and finally biopolitics. Yet Esposito problematises this possible reading: according to him, "Foucault refuses to respond conclusively" on how biopolitics and sovereignty relate to each other.[28] In *Society Must be Defended*, Foucault seems to believe that discipline is a "new mechanism of power," which is "the exact, point-for-point opposite of the mechanics of power" of sovereignty.[29] In the first volume of *History of Sexuality* he suggests that biopolitics is engendered by discipline the moment when the latter completely replaces the very grounds of sovereignty: "The ancient right to *take* life or *let* live was replaced by a power to *foster* life or *disallow* it to the point of death."[30] It is precisely in this definition of biopolitics as the power that both *fosters* and *disallows* life that Foucault's oscillations emerge. If fostering life – by protecting and enhancing it – is arguably very different from the sovereign idea of limiting life – through impositions, taxes, and so on – the possibility of *disallowing* life seems to converge with the sovereign's right of life and death over the subjects. This complication is already evident in *Society Must be Defended*, in which Foucault states not only that the new power is the exact "point-to-point" opposite of sovereignty but also:

> I wouldn't say exactly that sovereignty's old right – to take life or let live – was replaced, but it came to be complemented by a new right which does not erase the old right but which does penetrate it, permeate it.[31]

Esposito proposes to resolve this ambiguity, if not contradiction, by distinguishing between a "politics of life" (*politica della vita*) from a "politics over life" (*politica sulla vita*). The "politics of life" could be regarded as an affirmative biopolitics, which "fosters life," "in contrast to the commanding tendency of the sovereign regime."[32] While sovereignty "was exercised in terms of subtraction and extraction of goods, services, and blood," this affirmative biopolitics

> is addressed to the subjects' lives, not only in the sense of their defence, but also with regard to how to deploy, strengthen, and maximise life.[33]

The very fact that a "politics of life" focuses on managing life means that biopolitics as such entails the possibility of becoming a *thanatopolitics*, a "power of life [that] is exercised against life itself":[34] as Esposito writes, "biopolitics continually threaten[s] to be reversed into thanatopolitics."[35] Thanatopolitics, that is, a "politics *over* life" as, literally, a politics of death, corresponds to the "development of biopower and incremental growth in homicidal capacity,"[36] as exemplified in the deadly importance that Nazism granted to biology, eugenics and, broadly speaking, to the management of life. This aspect of biopolitics, or better, this inherent possibility of a *drift* of biopolitics towards thanatopolitics should not be regarded as a fracture with sovereignty but rather as a continuity with it, in that it is marked precisely by the return of the "ancient sovereign power of taking life";[37] biopolitics not only fosters life but can also "disallow" it.

It goes beyond the purpose of this article to solve Foucault's fluctuation between biopolitics and biopower, between a politics *of* life and a politics *over* life. Rather, it is important to remark that Di Vittorio's use of the term "biopolitical" seems to overlook this tension. His definition of "biopolitical psychiatry" does not bear any possible affirmative connotation; it is a completely negative perspective on biopolitics, and should rather be called a "thanatopolitical" psychiatry. Di Vittorio mainly refers the adjective "biopolitical" to the systematisation of the implementation of psychiatry at a state level, an example of which would be the recent *Programma di comunicazione contro il pregiudizio in salute mentale* promoted by Italy's Ministry of Health. After overflowing the walls of the asylum and becoming the generalised science of conduct and behaviour, psychiatry is turning into an explicit matter of interest for the state.

3.2 biopolitical psychiatry

According to Di Vittorio, biopolitical psychiatry adopts "an apparatus of prevention," which "classifies, pathologises, and institutionalises a much larger segment of the population."[38] The origins of biopolitical psychiatry can be traced back to the process that Foucault describes in his 1974–75 course *Abnormal*. By becoming a medico-legal expertise entrusted with the assessment of criminal behaviour, psychiatry also aimed at anticipating and preventing crime by detecting dangerousness in the everyday conduct of individuals. This potential dangerousness was identified with abnormality, and it was through this identification that psychiatry ceased to study exclusively the disorders of the mind in order to address the totality of the "inner space of the individual"; the latter became the "privileged object of the psychiatric gaze."[39] In other words, by establishing itself as the "science and technique of the management of anomalies,"[40] psychiatry evolved into a "general instance of the defence of society against the dangers that threaten it from within."[41]

Arguably, it is precisely this possible evolution that was overlooked by post-Basaglian psychiatrists. As Di Vittorio suggests, "reformed Italian psychiatry has failed to recognise that the new 'mental health' policy is bio-political."[42] Colucci agrees with Di Vittorio on this point, and, as he contends in his recent article "Scienza del pericolo, clinica del deficit," this is precisely the "biopolitical risk that many workers in the field of psychiatry run."[43] "Through medical emergency" these operators are asked to seize control over the sick person in order to enforce a social order that has not changed since the era of the asylum.[44] Despite Basaglia's reform, Italian psychiatry "could not completely put into brackets the medicalisation of suffering"; this ultimately amounts to continuing to regard all the needs of the psychiatric patient as symptoms of a mental illness.[45] Both Colucci and Di Vittorio seem sceptical towards the ultimate outcome of the psychiatric reform, which they deem as only partly successful, no matter how revolutionary it originally was.[46] While Rotelli had already anticipated that in Italy, as much as everywhere else, the field of psychiatry is still dominated by the medicalisation of madness,[47] Colucci adds that to date there are hardly any practices of assistance centred in the local community and administered by it.[48] On the contrary, psychiatric intervention increasingly focuses on a strictly medical approach that is state oriented, and the medicalisation of psychiatry is openly advocated as a necessary step, "after years of 'obscurantism' and 'ignorance'," towards the "restructuring of psychiatry into a scientific discipline."[49] Likewise, Di Vittorio observes that, after Basaglia, Italian psychiatry has "failed to understand how a 'good' mental health policy [. . .] can easily become the best alibi for a 'bad' mental health policy."[50] This "good" mental health policy is what Basaglia urged the most, "a clinical approach to the subject that entails curing his social bonds and the reconstruction of his belonging to a *polis*."[51] However, this "good" policy can easily turn into a "bad" policy that

> involves control and social normalisation through an apparatus of generalised prevention of pathology risks and the massive prescription of psycholeptic drugs.[52]

Biopolitical psychiatry is a consequence of the "generalised triumph of everything that has real or alleged neuroscientific validity";[53] it marks an unprecedented reduction of human nature to mere organic existence.

On the other hand, what remains of social and community intervention is still grounded on "models of rehabilitation," which is just another phrase for "techniques of adaptation to normality." The latter exploit "modules of training and assistance that have nothing to do with the specificity of the subject."[54] Even in reformed Italian psychiatry, the achievements of neurosciences mask the "old forms of protection of society against the madman."[55]

Basaglia had already anticipated this possible outcome of his reform. One year before his death, during a conference in Belo Horizonte on 17 November 1979, he stressed that the asylum is no longer "within the walls." Rather, it is to be found in our everyday life, since we are "medicalised and psychiatrised every time we go to see a GP."[56] Basaglia is adamant on this point: "We have now to act upon this new asylum."[57] From his perspective, it was urgent to "find a real content for this [. . .] alternative psychiatry,"[58] the ground on which to create a new psychiatry that does not re-propose the relations of power of the old disciplinary and institutional system; ultimately, a new psychiatry that does not become biopolitical in Di Vittorio's meaning of the term.

4 towards a clinic of lack

According to Colucci, Basaglia's proposal for an alternative psychiatry has been largely neglected, a fact that may have prompted the rise of a negatively biopolitical psychiatry as a replacement for the old disciplinary system. This proposal, in Colucci's interpretation, amounted to a clinic centred on the subject, the cure of his social bonds, and the reconstruction of the *polis*.[59] I believe that Colucci correctly identifies Basaglia's overall mission, but also overlooks two very problematic questions. That is to say, according to Basaglia, what is a subject? And also, what is the *polis* that should be reconstructed? Although to a certain extent I agree with Colucci, I also intend to problematise his position by suggesting that the psychiatric practice that Basaglia envisaged consisted, in brief, in the bracketing of mental illness as a preliminary step to a clinical approach based on the constitutional lack of subjectivity. Such an approach could be regarded as an affirmative biopolitical psychiatry, that is, a psychiatry that does not deny its biopolitical implications, but rather prevents them from turning into the application of a thanatopolitical biopower.

My notion of a "clinic of lack" is a response to Colucci's double assumption that Basaglia's clinical approach should be regarded both as a clinic of the subject, a "restitution of subjectivity" as he claimed in an earlier essay,[60] and a reconstruction of the *polis*. I will first problematise the notion of the Basaglian subject, and then the idea that a restitution of subjectivity could reconstruct a *polis*. In my opinion, Basaglia's thought should be contrasted and compared with Foucault's on both points.

4.1 subjects, individuals, utopias

According to Foucault, the individual is always-already woven into a network of relations of power, thus he regards him as an effect of power itself. This is clear, for instance, in *Psychiatric Power*, where Foucault suggests that the individual is nothing other than the result of certain "techniques of political power," such as "uninterrupted supervision, continual writing and potential punishment."[61] The process of individuation entails several other techniques, which Foucault describes elsewhere, such as examination,[62] expertise,[63] and normalisation.[64] In "The Subject and Power," Foucault adds that, throughout his work, his objective "has been to create a history of the different modes by which human beings are *made* subjects."[65] Arguably, in this statement Foucault proposes that human beings are not subjects, but they are *made* so. It could be advanced that, while the individual is the effect of power with respect to social relations, the subject is the effect of power with respect to the reflexive relation. That is to say, as human beings we are under an effect of power when we are in a relationship with others (individuation), but we are under the same effect also when we are alone with ourselves (subjectivity). Hence, we can assume that, according to Foucault, there is no subject who pre-exists the individual; the subject does not alienate himself only at the level of individuation. Human beings are always-already alienated in subjectivity, because the latter is itself the product of several "technologies of the self," which

> permit individuals to effect by their own means [. . .] a certain number of operations on their own bodies and souls, thoughts, conduct, and way of being, so as to transform themselves in order to attain a certain state of happiness, purity, wisdom, perfection, or immortality.[66]

Through these technologies subjects become capable of self-discipline and self-control.

Eventually there would be no need for someone to wield power, because subjects are themselves a product of self-disciplining techniques, thus they are always under a permanent effect of power. This was already clear to Foucault in *Discipline and Punish* (1975), in which he proposed that the ultimate aim of panopticism[67] was precisely to create a surveillance that is "permanent in its effects, even if it is discontinuous in its action."[68] Power no longer emanates from a centre, such as the sovereign or the supervisor. It is the subjects who exert power on themselves: instead of "domination over the object," the generalisation of panopticism inaugurates "[the object's] subjective participation in the act of domination."[69]

To this extent, Foucault's notions of "individual" and "subject" often seem to be overlapping. Although Foucault never clearly posits a complete convergence of individuality and subjectivity, from his writings it is not clear whether it is possible to trace a clear-cut distinction between the two notions or whether a subject *is*, all things considered, an individual. It seems to me that Foucault very often implies that human beings are always-already subjects (i.e., subjected to self-disciplining techniques) *and* individuals (i.e., individuated by a network of relations of power). Although we can distinguish theoretically between subjectivity and individuality to separate the effects of self-disciplining techniques from the effects of social relations of power, there is no such thing as a subject who is not an individual.

Despite the fact that Basaglia never developed a precise definition of the subject,[70] at first sight his late conclusions would seem to converge with Foucault's considerations on subjectivity. Yet it is crucial to remark outright that Basaglia's stance cannot be assimilated to Foucault's: Basaglia's subject does not correspond in any possible acceptation to an individual in Foucault's terms. At the beginning of his career, Basaglia referred to the subject using a Heideggerian-Binswangerian terminology: the subject is characterised by his *Dasein*, his being-in-the-world, his intrinsic capacity to create projects for his life, in short "the way in which he opens onto the world."[71] Yet if there is such a thing as a "Basaglian subject," it should not be regarded as something substantial, "with its most unassailable metaphysical connotations of unity, absoluteness, and interiority."[72] Since his early writings it is clear that for Basaglia the most important characteristic of human beings – the very definition of "human nature" – lies in the impossibility of their establishing a direct reflexive relationship with themselves without, at the same time, relating to the other. It is only in the "possibility of opposing himself to the external world" that the subject can "succeed in affirming himself."[73] This is Basaglia's notion of human nature: the latter cannot be regarded as purely biological; the organic aspect must be bracketed in order to avoid a priori theories that would impede the establishment of an intersubjective relationship with the patient. It is "only when man feels the necessity for a human relationship [that] he becomes such, inasmuch as he breaks his isolation to enter the world and give himself to it."[74] What we refer to as "psychic reality" is in itself constitutionally intersubjective, since

> when we refer to the term "psychic" we do not necessarily refer to something subjective and individual, because the individual partakes not only of himself, but also of all that surrounds him [...] that overcomes him and invests all other human entities, something inter-human of which every being partakes.[75]

Arguably, in this context, the only possible definition of human nature is to be found in a constitutional lack of subjectivity. Subjects cannot be such without relating to the other, in that

> man can only carry out an act of self-reflection through the gaze of the other; it is the gaze of the other as an intermediary that returns my self to me and makes me self-conscious.[76]

From a slightly different perspective, subjectivity in Basaglia's thought is to be considered mainly as a *utopian* construction. According to him, utopias are those ideas that help shape reality,[77] which he encouraged considering as a "practical truth."[78] Reality should not be regarded as something given and static, but rather as a dynamic entity, which is shaped by the actual needs of people. Utopias are a "foreshadowing" of this reality,[79] as they are "a constant search to meet needs and construct a life that is possible for all mankind."[80] Reality should be constantly shaped and reshaped in view of universalism, following this continuous search based on the needs of human beings. If reality is to be seen as an unstable condition, then utopia is the element that prevents it from becoming static. I would go so far as to suggest that utopias produce what Esposito defines as a "porous democracy"[81] – which is in itself an appropriate expression to describe an affirmative biopolitics – "whose forms are always the object of innovation and self-overcoming."[82]

In the specificity of psychiatry, "utopia is that daily practice that enables us to overthrow science [and] technique."[83] The utopia at which psychiatric therapy should aim consists precisely in approaching as closely as possible the subjectivity of the patient. In short, we must suppose that there is a subject that pre-exists the relations of power, even if we are dealing with a constitutionally lacking subject. The closer therapy gets to this lacking subject, the more it is able to diminish the alienating effects of power, that is, to recover a human being that is not reduced to the effects of power – which are imposed both from above and from within the subject. Necessarily, the human being that is recuperated in this way is not isolated: by accepting his constitutional lack, the subject reveals that his nature is to be in constant and radical need of the other.

Even if there is no proper "way out" of the individual, even if it is not possible to "empty [the subject] out, free [him] from himself,"[84] according to Basaglia, psychiatry should keep aiming at "returning subjectivity" (*restituir[e] la soggettività*) to the mentally ill.[85] In short, this is what I propose to call a "clinic of lack."[86] The individual may be the effect of relations of power, but for Basaglia the subject as such cannot be reduced to them: it is in this crucial point that Basaglia's stance radically differs from Foucault's. Through the bracketing of mental illness it seems somehow possible to recuperate a pre-existing subjectivity – even if it amounts only to a constitutional lack, the need for the other.

4.2 communitas *and* immunitas

Having said this, it is now time to focus on my second problematisation of Colucci's claim: what is the *polis* that should be reconstructed through the restitution of subjectivity? Instead of developing a clinical approach grounded on the constitutional lack of the subject, post-Basaglian psychiatry developed into a clinical approach centred on a unitary conception of the individual. Such an approach overlooks the constitutional lack of the subject, with the aim of rehabilitating the mentally ill and reintegrating them into society. That is to say, it ultimately seeks to reintroduce the individual into the place where he is best woven into those relations of power that can keep him under control. Even social psychiatry, with its community approach, aims at reintegrating and rehabilitating the patient, which, according to Basaglia, is a mere chimera in that this process does not properly heal him, but prepares for "his return into the productive circle."[87] Social psychiatry does not reintegrate the subject into a "community" but into the capitalist productive system, thus perpetuating the aim of disciplinary psychiatry through different means. Could the reconstruction of the *polis* be limited to the reintegration of the sick person into society, or should we rather aspire to the creation of a community?

Community is a critical keyword in contemporary psychiatry, in terms of which this "overused" concept normally has a very straightforward meaning.[88] According to Thornicroft and Tansella, the definition of community can readily be drawn from the *Concise Oxford Dictionary*: as far as psychiatric intervention and mental health care are concerned, the community is

> a defined population, for whom an integrated system of mental health care can be provided. Such a population may be geographically defined or may be identified by some other key criteria.[89]

If we accepted this acceptation of community, returning to the subject in order to reconstruct a *polis* would amount to a vicious circle. By bracketing mental illness and returning the patient to his constitutionally lacking subjectivity, the psychiatrist's ultimate ambition would still be to assimilate the patient into a community defined as a population.[90] This is, arguably, no more than a space regulated by the usual relations of power that would eventually demand that the reintegrated patient alienate himself once again as a docile element of the community.

On the contrary, I believe that resuming a constitutionally lacking subjectivity in order to reconstruct a *polis* entails a wider conception of community, which I derive from Esposito's analysis of the *communitas*. As Esposito puts it, the community is neither "a collective subject" nor a "totality of subjects"; rather, it amounts to "the relation that makes them no longer individual subjects."[91] In other words, it is a "no-thing" (*ni-ente*),[92] a "nothingness" that subtracts the subject from the "identity with himself" and delivers him to an "irreducible otherness."[93] Contemporary individuals tend to protect themselves from belonging to this kind of community, as this would entail a certain loss of their identity:

> modern individuals truly become such [. . .] if they are released from, exonerated, or relieved of that contact which threatens their identity, exposing them to possible conflict with their neighbour, exposing them to the contagion of the relation with others.[94]

Individuals tend to avoid the risk of losing their identity and belonging to an actual community. Understood in these terms, the notion of community does not amount to a

> multiplication of subjectivity for an indeterminate number of individuals, just as the individual would constitute a fragment of community that is simply waiting to have a relation with others.[95]

On the contrary, this kind of community is composed of human beings who understand their constitutional lack, their constant need for the other, and accept it: they accept the risk they run of losing their individuality.

According to Esposito, the discrimination between affirmative biopolitics and its negative counterpart – thanatopolitics/biopower – is based on the distinction between community and what seems to be its exact opposite, that is to say, immunity. Proceeding from the etymology of the word "community," composed of *cum-*, "what is *not* proper,"[96] and *munus*, the gift that "one *must* give and [. . .] one *cannot not* give,"[97] Esposito advances the view that community is not based on a common property but on a debt; community "does not add something but *subtracts*" (*non dà un più, ma dà un "meno"*).[98] In the community, individuals are precisely expropriated "of their initial property [. . .] namely, their very subjectivity."[99] *Immunitas* is the exact opposite of the community: it is the avoidance of the *munus*, the debt or lack that constitutes those who belong to a community. On the one hand, *immunitas* is the "original autonomy or a subsequent release from a previously incurred debt."[100] Those who are immune are exempted from the *munus* that would introduce them to the community. On the other hand, from a biomedical perspective, it is also the "condition of refractoriness of the organism towards the danger of contracting an illness."[101] *Immunitas* is a process of reaction to the external world, bacteria, and other possible vectors of contagion. Yet in so far as vaccinations themselves entail the inoculation of a small amount of the "evil" they counteract, immunity *cannot* be regarded as the exact opposite of community. There is no community without a certain amount of immunity. Just like the human body, the community needs some level of protection and immunisation: this is a precondition of affirmative biopolitics. To follow the organic metaphor, it is only when there is an excess of immunisation that a reversal in the function of antibodies takes place: the organism develops an auto-immune syndrome, such as lupus. The body fights against itself, since "immunity is necessary to protect our life, but when it exceeds a certain threshold, it ends up disallowing it."[102] This is the opposite of an affirmative biopolitics; it becomes thanatopolitics, an exertion of ruthless biopower.

In short, immunisation is what grounds the very process of individuation: the individual immunises himself from his constitutional lack and from the "dissolutive possibility" of community.[103] Whilst this is, to a certain extent, a necessary step in the formation of a community, an excess of immunisation arrests the community, whereby the latter is perceived as a threat to subjectivity qua the individual, as the ultimate limit to its wholeness.

What I propose to call a "clinic of lack" seems to aim precisely at a partial refusal of immunisation, and a return to a community understood in Esposito's terms. The reintegration into a community entails for the subject an acceptance of his constitutional lack, the acknowledgement that he "is characterised by an original lack that cannot be filled in":[104] this lack is nothing other than the need for the other.

Through psychiatric therapy, understood in Basaglia's terms, the patient does not achieve a substantial subjectivity, the belief that he is unique and detached from the world: this would only reinforce his immunisation and would impair the formation of a *communitas*. On the contrary, the patient should discover and accept his constitutional lack, the need to constantly refer to the other in order to understand himself, that is, the need to refer to the other in order to define his own subjectivity.

Such a "clinic of lack" necessarily involves the acceptance of man's constitutional lack, an idea that Basaglia formulated as early as 1953 in "Il mondo dell'incomprensibile schizofrenico," where he claimed that human beings cannot establish a direct relationship with their own self if they are not in a relationship with the other. Basaglia's primary concern in his early work was to define a method, which he sought in Binswanger's *Daseinsanalyse* and in phenomenological psychiatry, capable of accounting for both the psychiatrist's and the patient's subjectivities, as opposed to the positivist subject–object relationship. It could be said that Basaglia's early conception of the relationship between the patient and the psychiatrist is the prototype of this idea of community: both psychiatrist and patient call into question their own subjectivities and accept their constitutional lack, their need for the other. The psychiatrist eventually should accept to lose, as far as this is possible, his immunisation: bracketing mental illness, establishing an intersubjective relationship, entering into it "unarmed,"[105] refusing the authoritarian role of the psychiatrist, in short, all the conditions of the "ideological void" opened by the implementation of Law 180 should be read as a practical attempt at reducing the paradigms of immunisation in psychiatry and at establishing a *communitas*; as the only means to define a renewed therapeutic approach.

To sum up, if there is such a thing as a "real content" of Basaglia's "alternative psychiatry" it seems to me that it amounts to a therapy that aims at returning the subject to a community, understood in Esposito's terms, rather than at a reintegration into society. In other words, such therapy is guided by the utopian aim of helping the patient understand and accept his constitutional lack, so that he may partake of a (utopian) community of human beings who have in turn accepted their lack and their need for the other. This utopian community is not an ideal world "where everybody is good, where all relationships are marked by the deepest of humanisms, where work is always rewarding,"[106] but a community where the paradigms of immunisation are kept to that minimum which guarantees its survival, and thus do not run the risk of turning it into a thanatopolitical biopower.

Because of all this, it is possible to infer not only that Basaglia was aware of the possible "negatively biopolitical" outcomes of his reform, the thanatopolitical results of biopolitics in Esposito's terms, but also that he proposed an affirmative biopolitical psychiatry *ante litteram*. Contrary to what Di Vittorio claims, biopolitical psychiatry cannot be regarded as entirely negative, and Basaglia anticipated this. It is impossible and maybe not even desirable to avoid or counter biopolitics, renouncing completely the paradigms of immunisation. Rather, what Basaglia seemed to suggest is that it is possible and urgently necessary to envision a biopolitical psychiatry that does not turn into its thanatopolitical counterpart; an affirmative biopolitical psychiatry that moves from the premise that a clinic of lack would reduce the paradigms of immunisation.

notes

1 Literally, "mettere tra parentesi la malattia mentale" means "putting mental illness into brackets" or, more accurately, "into parentheses." However, the Italian phrase "tra parentesi" also means "incidentally," so "mettere tra parentesi la malattia mentale" also implies "treating the issue of mental illness as incidental to other issues." The phrase has been translated as "bracketing mental illness" but this is somewhat misleading; although "to bracket" can mean "to put into brackets," i.e., "to segregate" or, figuratively, "to set aside," it can also mean "to couple or join with a brace." As "bracketing" is a convenient shorthand

expression for what would otherwise be a lengthy circumlocution, it will be used in this article in inverted commas to indicate its special meaning in the context of Basaglia's thought.

2 M. Colucci and P. Di Vittorio, *Franco Basaglia* (Milan: Mondadori, 2001) 304.

3 As Revel has it, affirmative biopolitics

> is the need to think biopolitics also within the framework of an affirmation of being, as a radical positivity. [Affirmative biopolitics does] not exclude the creative invention of forms of being, or ways of life, that is, an expressive capacity or power to act that exceeds the relations constructed by *dispositifs* of power. (J. Revel, "Identity, Nature, Life: Three Biopolitical Deconstructions," *Theory, Culture and Society* 26 (2009): 52)

As far as psychiatry is concerned, I will try to illustrate in brief this oscillation between bio- and thanatopolitics through the example of psychopharmacology. We could argue that, in an affirmative biopolitical context, it would be impossible not to rely on pharmacology to treat mental illnesses and even milder psychological ailments, such as regular stress, occasional insomnia, etc. An appropriate use of psychopharmacology undoubtedly improves the conditions of life for many people. Yet an uncontrolled use of psychopharmacology easily degenerates into what could be defined as a thanatopolitical psychiatry, grounded in the reduction of all human emotions and states of being to chemical reactions, which can be enhanced, modified and restored to a "normal" state in case an imbalance such as a mental illness sets in, i.e., what Rose defines as the *neurochemical self* (N. Rose, "Neurochemical Selves," *Society* 41 (2003): 46–59).

4 This uncommon use of the term "clinic" is borrowed from the English translation of Foucault's *La Naissance de la clinique* (M. Foucault, *The Birth of the Clinic*, trans. A. Sheridan (London: Routledge, 2008)). Sheridan explains a duality of meaning in the French term which does not exist in English: when Foucault speaks of *la clinique* he is thinking of both clinical medicine and the teaching hospital. So if one wishes to retain the unity of the concept one is obliged to use the rather odd-sounding "clinic" (Foucault, *The Birth of the Clinic* vii). It is in the sense of "clinical medicine" that I am here using the term "clinic."

5 "Scienza del pericolo," *aut aut* 349 (Oct.–Dec. 2008): 115.

6 For two insightful intellectual biographies of Franco Basaglia, I refer the reader to Colucci and Di Vittorio (M. Colucci and P. Di Vittorio, *Franco Basaglia* (Milan: Mondadori, 2001)), and to Parmegiani and Zanetti (F. Parmegiani and M. Zanetti, *Basaglia, una biografia* (Trieste: Lint, 2007)).

7 F. Basaglia, "La distruzione dell'ospedale psichiatrico come luogo di istituzionalizzazione" [1964] in *L'utopia della realtà*, ed. M. Giannichedda (Turin: Einaudi, 2005) 17–26.

8 *Franco Basaglia* 299.

9 F. Basaglia, "Conversazione: a proposito della nuova Legge 180" [1980] in *Scritti*, ed. F. Ongaro Basaglia (Turin: Einaudi, 1981) 2: 479.

10 M. Foucault, *History of Madness*, trans. J. Murphy and J. Khalfa (London: Routledge, 2005) 44.

11 Ibid. 249.

12 "Conversazione" 470.

13 M. Donnelly, *The Politics of Mental Health in Italy* (London: Routledge, 1992) 100.

14 F. Basaglia, "Prefazione a *Il giardino dei gelsi*" [1979] in *Scritti* 2: 472.

15 Ibid.

16 Social psychiatry is a generic definition given to those approaches in psychiatry that are based on the principle that mental illness is not exclusively an organic disease. Social psychiatry focuses on the impact of society on the aetiology of mental illness and experiments with therapeutic approaches based on the reintegration of the ill in their community. It gained influence during the late 1930s, especially through the works of Karen Horney and Erik Erikson.

17 "Prefazione a *Il giardino dei gelsi*" 472.

18 F. Basaglia, "Introduzione generale ed esposizione riassuntiva dei vari gruppi di lavori" [1981] in *Scritti*, ed. F. Ongaro Basaglia (Turin: Einaudi, 1981) 1: xxii.

19 F. Basaglia, "Introduzione ad 'Asylums'" [1969] in *Scritti* 2: 35–36.

20 "Prefazione" 472.

21 Ibid. 133.

22 P. Di Vittorio, "From Psychiatry to Bio-politics Or the Birth of the Bio-security State" in *Michel Foucault and Power Today: International Multidisciplinary Studies in the History of the Present*, eds. A. Beaulieu and D. Gabbard (Oxford: Lexington, 2006) 79.

23 M. Foucault, *Society Must be Defended: Lectures at the Collège de France, 1975–76*, trans. D. Macey (New York: Picador, 2003) 242.

24 Ibid.

25 Foucault has never clarified exhaustively the distinction between biopower and biopolitics. As Esposito has it, Foucault "never sufficiently articulated the concept of politics – to the point of substantially superimposing the expressions of 'biopower' and 'biopolitics'" (R. Esposito, *Bíos: Biopolitics and Philosophy*, trans. T. Campbell (Minneapolis: U of Minnesota P, 2008) 44). Rabinow and Rose tried to resolve this ambiguity by defining biopower(s) as the power exerted on the population, when this is understood as a collective of bodies. On the other hand, they define biopolitics as the organisation and systematisation of such power by the political forces that govern these collectives. Nevertheless, Rabinow and Rose also agree that Foucault is "somewhat imprecise in his use of terms" and they regret that there has been no systematic study of his "sketchy suggestions" on biopolitics and biopower (P. Rabinow and N. Rose, "Biopower Today," *Biosciences* 1 (2006): 197).

26 *Society Must be Defended* 242.

27 Ibid. 243

28 *Bíos* 34.

29 *Society Must be Defended* 35–36.

30 M. Foucault, *The History of Sexuality, Volume I: The Will to Knowledge*, trans. R. Hurley (London: Penguin, 1998) 138.

31 *Society Must be Defended* 241.

32 *Bíos* 37.

33 Ibid. 36.

34 Ibid. 39.

35 Ibid.

36 Ibid.

37 Ibid. 41.

38 "From Psychiatry to Bio-politics" 73.

39 N. Rose, *The Politics of Life Itself: Biomedicine, Power, and Subjectivity in the Twenty-First Century* (Princeton: Princeton UP, 2007) 194.

40 M. Bertani, "La nascita della psichiatria dallo spirito della follia. Nota storica su 'il potere psichiatrico'," *aut aut* 323 (2004): 61.

41 The most alarming of these dangers is *idleness*. As Basaglia remarked, the scientific pretensions of institutional psychiatry were invalidated by its economic bias. Basaglia maintains that, in a capitalist society, the distinction between normal and abnormal, as well as that between the sane and the mentally ill person, is based on the criterion of productivity. Basaglia substantiates his argument with a statistical study that he conducted along with his team in the asylums of Trieste and Volterra. The data showed that the number of hospitalisations and discharges followed the general state of the economy. That is to say, higher hospitalisation rates corresponded to periods of economic recession, whereas a higher rate of discharges matched with periods of economic development. It seemed as though,

> depending on whether a society is undergoing a time of development or one of recession and crisis, we can witness the corresponding enlargement or shrinking of the limits imposed by norms and, hence, the dilation or contraction of tolerance towards abnormal behaviour. (F. Basaglia, "La giustizia che non riesce a difendere se stessa" [1976] in *Scritti* 2: 386)

42 "From Psychiatry to Bio-politics" 75.

43 "Scienza del pericolo" 113.

44 Ibid.

45 Ibid. 111–12.

46 Foucault shared similar reservations concerning the effectiveness of reformism in psychiatry, claiming that it brought about a renewed medicalisation of madness and new relations of power in the territorial centres which replaced the asylum. See Foucault, *Psychiatric Power* 342ff. and "L'Asile illimité" in *Dits et écrits* (Paris: Gallimard, 1994) 271–75.

47 F. Rotelli, "Foucault a Trieste," *Rivista Sperimentale di Freniatria* 3 (2005): 39.

48 "Scienza del pericolo" 118.

49 Ibid. 121.

50 "From Psychiatry to Bio-politics" 75.

51 "Scienza del pericolo" 115.

52 Ibid. 75. The massive use of psychopharmacology is one of the most prominent consequences of the reduction of mental processes to organic (brain) activity. As Holmer Nadesan has it, psychopharmacology is based on the assumption that "mental states [are] epiphenomena of brain states and that chemical imbalances in the brain produce mental imbalances" (M. Holmer Nadesan, *Governamentality, Biopower and Everyday Life* (London: Routledge, 2008) 168; see also 162ff.). An insight into the relationship between psychopharmacology and biopolitics can be found in P. Bourgois, "Disciplining Addictions: The Bio-politics of Methadone and Heroin in the United States," *Culture, Medicine and Psychiatry* 24 (2000): 165–95. Bourgois claims that methadone therapy as a replacement for heroine addiction has been used as a means to control unproductive individuals, while the side-effects of the treatment are often underemphasised.

53 "Scienza del pericolo" 115.

54 Ibid.

55 Ibid.

56 F. Basaglia, *Conferenze brasiliane* (Milan: Cortina Raffaello, 2000) 176.

57 Ibid. 181.

58 Ibid. 167.

59 "Scienza del pericolo" 115

60 M. Colucci, "Dissipazione" in *Follia e paradosso. Seminari sul pensiero di Basaglia* (Trieste: Edizioni E, 1995) 92.

61 M. Foucault, *Psychiatric Power: Lectures at the Collège de France, 1973–74*, trans. G. Burchell (New York: Palgrave, 2006) 56.

62 Idem, *Discipline and Punish: The Birth of the Prison*, trans. A. Sheridan (London: Penguin, 1991) 184–92.

63 M. Foucault, *Abnormal: Lectures at the Collège de France, 1974–1975*, trans. G. Burchell (New York: Picador, 2003) 1–30.

64 *Discipline and Punish* 182–83.

65 M. Foucault, "The Subject and Power" in *Michel Foucault: Beyond Structuralism and Hermeneutics*, eds. H.L. Dreyfus and P. Rabinow (Brighton: Harvester, 1982) 208.

66 M. Foucault, "Technologies of the Self" in *Technologies of the Self: A Seminar with Michel Foucault*, eds. L.H. Martin, H. Gutman, and P.H. Hutton (London: Tavistock, 1988) 18.

67 The concept of "panopticism" originates from Jeremy Bentham's utopian prison, the *Panopticon*. This was a circular structure that had the inmates' cells on the external perimeter, while the superintendent was located at the centre. He could see all that was happening in the cells, while the inmates never saw the central surveillance structure. As Foucault has it, the *Panopticon* induces "in the inmate a state of conscious and permanent visibility that assures the automatic functioning of power" (*Discipline and Punish* 201). The *Panopticon* amounts to the utopia of disciplinary power: achieving the total visibility of all individuals, without exposing the source of power.

68 *Discipline and Punish* 201.

69 *Bíos* 35. This is why, according to Rovatti, "for Foucault there is no subject, the subject is an invention full of negative consequences, consequences that can even be destructive" (P.A. Rovatti, "Il soggetto che non c'è" in *Foucault, oggi*, ed. M. Galzigna (Milan: Feltrinelli, 2008) 217), because, as Seigel has it, "sought in the name of freedom, such subjectivity opened individuals to domination by the powers" (ibid. 276). There is no such thing as an independent subject, as the very way we perceive ourselves is always-already determined by relations of power.

70 Basaglia is often inconsistent in his use of philosophical vocabulary. In his early works, it is clear that whenever he mentions the subject or subjectivity he is using these notions in a phenomenological way in opposition to the positivist conception of objectivity. After 1964 and the "political turn," his language becomes less specific. He often uses interchangeably the terms *soggetto* (subject), *individuo* (individual) and *persona* (person). Hence, the following discussion is to be considered as an analysis of the concept of subject as it is *implicit* in Basaglia's writings, rather than as a study of his own definition of the subject.

71 F. Basaglia, "Il mondo dell'"incomprensibile' schizofrenico attraverso la 'daseinsanalyse.' Presentazione di un caso clinico" [1953] in *Scritti* I: 5.

72 R. Esposito, *Communitas: The Origin and Destiny of Community*, trans. T. Campbell (Stanford: Stanford UP, 2010) 1.

73 F. Basaglia, "Il corpo nell'ipocondria e nella depersonalizzazione. La coscienza del corpo e il sentimento di esistenza corporea nella depersonalizzazione somatopsichica" [1956] in *Scritti* I: 171.

74 F. Basaglia, "Su alcuni aspetti della moderna psicoterapia: analisi fenomenologica dell'"incontro'" [1954] in *Scritti* I: 35.

75 Ibid. 43.

76 F. Basaglia, "Corpo, sguardo e silenzio. L'enigma della soggettività in psichiatria" [1965] in *L'utopia della realtà*, ed. M. Giannichedda (Turin: Einaudi, 2005) 32. The "look" is one of the fundamental forms of interaction between human beings, since the very beginning of psychical ontogenesis. Sartre remarks that "the Other's look, as the necessary condition for my objectivity, is the destruction of all objectivity for me" (J.-P. Sartre, *Being and Nothingness: An Essay on Phenomenological Ontology*, trans. H.E. Branes (London: Methuen, 2008) 269). Foucault defines the gaze as an instrument of power. This is the case above all in disciplinary power, which requires continuous control over the individual's life, time and especially body. In Foucault's words, "in the disciplinary system, one is not available for someone's possible use, one is perpetually under someone's gaze, or, at any rate, in the situation of being observed" (*Psychiatric Power* 47). In "Logical Time and the Assertion of Anticipated Certitude," Lacan identifies in the gaze the fundamental structure of intersubjectivity (J. Lacan, "Logical Time and the Assertion of Anticipated Certainty" in *Écrits: The First Complete Edition in English*, trans. B. Fink (London: Norton: 2007) 161–75).

77 F. Basaglia, "Peacetime Crimes: Technicians of Practical Knowledge" in *Psychiatry Inside Out: Selected Writings of Franco Basaglia*, eds. A.M. Lovell and N. Scheper-Hughes; trans. A.M. Lovell and T. Shtob (New York: Columbia UP, 1987) 157.

78 Ibid.

79 Ibid.

80 Ibid.

81 R. Esposito, "Prefazione" in *L'impersonale. In dialogo con Roberto Esposito*, ed. L. Bazzicalupo (Milan: Mimesis, 2008) 26.

82 Ibid.

83 *Franco Basaglia* 74.

84 "Il soggetto che non c'è" 219.

85 "Dissipazione" 92.

86 It is crucial to elucidate the notion of "clinic of lack," which was introduced as *clinica della mancanza* by Massimo Recalcati, one of the most renowned Lacanian psychoanalysts in Italy. Despite the resemblances that may emerge, in this article I am not referring to Recalcati's *clinica della mancanza*, which amounts to a psychoanalytical approach to unconscious desire understood as originating from a constitutional lack in the subject (M. Recalcati, *Clinica Del Vuoto. Anoressie, Dipendenze, Psicosi* (Milan: Angeli, 2002)).

87 *Conferenze brasiliane* 123.

88 E.D. Acheson, "That Over-Used Word Community," *HealthTrends* 17 (1985): 3.

89 M. Tansella and G. Thornicroft, "Planning and Providing Mental Health Services for a Community" in *New Oxford Textbook of Psychiatry*, eds. M. Gelder, N. Andreasen, J. Lopez-Ibor, and J. Geddes (Oxford: Oxford UP, 2009).

90 The notion of population is crucial to understanding Foucault's conceptualisation of biopolitics, as it "made possible a logic in which the government of the state came to involve [both] individualisation and totalisation." Defining the notion of "population" enables Foucault to articulate the "development of anatomo-political techniques aimed at the individual body [with] the development of bio-political techniques aimed at the collective or social body" (B. Curtis, "Foucault on Governamentality and Population: The Impossible Discovery," *Canadian Journal of Sociology* 27 (2002): 506). As Curtis continues, "population" entails the definition of an essence common to all its members, which can only be articulated by statistical means (ibid. 510). In the context of psychiatry (and, broadly speaking, medicine), defining a population – which Tansella and Thornicroft seem to regard as a synonym for community – means reducing all its members to their shared organic traits, that is, to their human nature

understood as a set of biological and universal features. In light of these considerations, it is all the more important to understand that here the notion of individual does not logically precede that of population, but is one of its consequences. The traits and features of each single individual depend upon the statistical description of the population to which it belongs.

91 *Communitas* 139.

92 *Niente* in Italian means both "nothing" and "nothingness." *Ente* means "entity"; from this the pun *ni-ente* follows, that is, an entity which is as such nothing, yet is positively composed by the communal sharing of its nothingness.

93 *Bíos* 81.

94 *Communitas* 13; translation modified.

95 Ibid. 73.

96 Ibid. 3; emphasis in original.

97 Ibid. 5; emphasis in original.

98 Ibid. 6; my emphasis; translation modified.

99 Ibid. 7.

100 R. Esposito, *Immunitas. Protezione e negazione della vita* (Turin: Einaudi, 2002) 8.

101 Ibid. 9.

102 R. Esposito, "Biopolitica, immunità, comunità" in *Biopolitica. Storia ed attualità di un concetto*, ed. A. Cutro (Verona: Ombre Corte, 2005) 161.

103 *Immunitas* 18.

104 R. Esposito, *Terza persona. Politica della vita e filosofia dell'impersonale* (Turin: Einaudi, 2007) 135.

105 "Prefazione a *Il giardino dei gelsi*" 472.

106 F. Basaglia, ed., *Che cos'è la psichiatria?* (Milan: Baldini Castoldi Dalai, 1997) 20.

lorenzo chiesa

THE BIO-THEO-POLITICS OF BIRTH

I the sacredness of birth

In his controversial newspaper articles, interviews, and letters of 1975 opposing the legalisation of abortion in Italy, Pier Paolo Pasolini often qualifies life as sacred. Problematically, this is, for him, in the first instance, an "obvious" fact that does not need to be demonstrated and must be taken for granted.[1] Pasolini thus refuses to conceive the notion of life (or even nature) as a merely symbolic construction: before the emergence of both religion and law, life has in itself an inherent meaningful value and is as such hieratic.[2] Yet, at the same time, he also readily emphasises the political implications of his claim. According to him, we are dealing here with a "principle" that is "even stronger than the principle of democracy."[3] In a few words, the sacredness of life allows us to think politics straightforwardly as an "ecology." The first and most important biopolitical objective of the human species has always been its survival and propagation; from this perspective, human birth is sacred and "blessed" (*benedetta*) as long as it remains a "guarantee" for the life of the species, "the continuation of man."[4] Conversely, when births far outnumber deaths and we face a demographic catastrophe, every new child starts to represent a "contribution to the self-destruction of humanity, and is thus damned [*maledetto*]."[5] Given that today we have reached such a threshold, abortion could eventually be legalised only on the basis of ecological "extenuating circumstances."[6]

Interestingly enough, in this context, Pasolini shows himself to be perfectly aware of the semantic complexity inherent to the Latin term *sacer* – "in our world, life is no longer sacred except in the sense of damned (*sacer* has both meanings)"[7] – that is so crucial to Giorgio Agamben's reflection. In addition to this, he considers the positive acceptation of *sacer* – i.e., blessed birth – to be inextricable from an "anthropological world of poverty"[8] which, as he specifies elsewhere, is not to be equated with material destitution (as a matter of fact, in vast areas of the planet indigence coexists with a demographic explosion) but rather amounts to the "fundamental quality of the people" as opposed to those who exploit them, or also, "a certain simplicity" as pure precarious resistance.[9] Juxtaposing his remarks about the damned character of birth as it stands at present with his resumption of the political significance of poverty, we can understand why he might laconically conclude that "superfluous goods render life superfluous."[10]

While, at this stage, it should already be clear that the debt of radical contemporary Italian thought to Pasolini is considerable – especially with regard to its treatment of bare life and the emancipatory potential of the pauper – the true significance of his contribution to a philosophy of biopolitics fully emerges only when we focus more closely on the link he establishes between sacred birth and *economic* reproduction. As Antonio Negri has argued convincingly in "The Italian Difference," philosophy seems to be possible today only in the guise of a biopolitical investigation into the theme of reproduction, sexual and economic.[11] Beyond Agamben, whose overall analysis of *homo sacer* underestimates the importance of the reproductive dimensions of a politics of and on life, the late Pasolini's critique of consumerism is held together by one common thread: exploitative biopolitics is that which directly short-circuits biology with economy, sexuality with Capital, through a diligent management of reproduction.[12] This corresponds to a two-layered ideological process. On a first level, the new consumerist power has replaced the sacrality of birth, compromised by an impending Malthusian crisis, with "the sacrality of consumption as a rite [. . .] and of commodities as a fetish."[13] On a second level, such de-sacralisation (or "consumeristic secularism")[14] – which also amounts to a form of re-sacralisation – is itself supported by the imposition of a model of sexual freedom that "includes the consumption of all the surplus considered necessary by a modern couple."[15] In other words, in promoting sexual emancipation, the falsely progressive ideology of neo-capitalism is also enjoining us to consume the economic surplus:[16] sexual emancipation enhances consumption, and ideological hedonism thus represents first and foremost a way of controlling the economy biopolitically.[17] Or, more bluntly, the "socially erotomaniac" model of the emancipated couple corresponds in the sphere of the reification of all human relations precisely to nothing other than what "a car is for an average consumer," namely, "a proof of a normal existence."[18]

These considerations should throw new light on Pasolini's much-deprecated opposition to the legalisation of abortion. It would not do him justice to reduce his unorthodox arguments to a doxastic Catholic pro-life stance. On the other hand, it is precisely his unmediated engagement in the "pro-life/pro-choice" debate that may indicate how Agamben's understanding of sacred life can become more intelligible politically only when he will openly measure it against the background of current affairs. While his notion of biopolitics has been applied hastily to the most disparate cultural contexts, we could suggest that it is perhaps these simplifying generalisations that account for the Library of Congress's misleading – but also highly symptomatic – cataloguing of the English translation of *Homo Sacer* under the "right to life" heading . . .[19]

We have seen how Pasolini's biopolitical analysis of abortion focuses on the economic value of sexuality in our apparently liberal consumer society. More precisely, Pasolini emphasises that the debate on abortion should first and foremost be a debate on coitus: as he points out, "when people speak about abortion they omit to speak of what logically precedes it, that is, coitus," and yet "coitus is political."[20] Today's "system of pseudo-tolerance" turns copulation into a "consumerist duty" and therefore into an "institutional act."[21] In other words, we need to bear in mind that, in our world, those in power can achieve the enhancement of economic reproduction and, at the same time, attempt to control overpopulation only by means of an enhancement of sexual copulation that excludes sexual reproduction. What in the end generates real surplus value is, for Pasolini, the subtraction of birth from ideologised coitus:

> Legalised abortion is as a matter of fact – there is no doubt about this – a great convenience for the majority of people. Above all because it would make coitus – heterosexual copulation – even easier and there would be no more obstacles to it. But who did tacitly want [. . .] this freedom of the coitus of the "couple" [. . .] who did tacitly promulgate it and make it enter our habits, in a way that is by now irreversible? The consumerist power, the new fascism.[22]

Three important and interrelated corollaries follow from this argument:

1. Contemporary biopolitics "ontologises" coitus in such a radical way that it ultimately "annuls" it. In other words, the "offensive naturalness" with which we are asked to relate to copulation eliminates sexual difference understood as symbolic difference and, in "de-sentimentalising" or "de-sacralising" man, it animalises him.[23]

2. Contemporary biopolitics dismantles the political. The people's ideological acceptance of coitus as an institutional act, that is, their experiencing it as "an irresponsible mania," coincides with their "desertion of politics." Conversely, the constituted power's legalisation of abortion relies on an unprecedented "political illegality"; the destruction of any "social equilibrium" brought about by the brutality of consumerist neo-capitalism is being contradictorily both countered and nourished by an effort to regulate, "and even reinforce in a time of crisis, the [apolitical] social *status quo*."[24]

3. Contemporary biopolitics is not able to exercise its power on couples who produce offspring [*prole*] as long as they remain, by definition, *prole-tarian.* Consequently, all proletarian couples must culturally be transformed into petit-bourgeois consumer couples.[25]

With regard to this last point, it is crucial to stress that, according to Pasolini, the "neo-secular hedonism"[26] of consumerist capitalism produces a new kind of apolitical exploitative social relation between the classes that is unchangeable. Any form of social alterity, the very possibility of inventing a different set of social relations, is excluded once and for all.[27] In other words, the obliteration of the popular *culture* of the lower classes – ultimately based on sacred life as blessed birth – is far from corresponding to an overcoming of their unfavourable *economy*. Rather, such a process increases the permanent alienation of their labour precisely in that it imposes a consumerist cultural model and an "alienation of love."[28] In so far as consumerism is cynically entwined with a false realisation of hedonistic "civil rights" for all, class struggle is foreclosed.[29]

Again, these specifications should prevent us from hurriedly condemning Pasolini's opposition to abortion as simply reactionary. Although he himself admits that "in this nostalgia for an idealised sacredness that possibly never existed, for sacredness has always been institutionalised, [...] there is something wrong, irrational, and traditionalist" – something that "contradict[s] my Marxism [...] what Hegelianism has taught me about movement [and] dialectics"[30] – on closer inspection even his a priori assumption about the sacred status of life as such seems to lend itself to alternative interpretations. While confirming the importance of the foetus's "furious, total, and essential will to life" – and the fact that its longing to "realise its potentiality" is something "absolute"[31] – Pasolini also observes that defending life unreservedly, as a matter of sheer principle, is a form of "pure *qualunquismo* [anything goes]" – that is, radical political apathy.[32] Above all, he is eager to specify that his notion of sacred life, which generated a series of negative reactions from Italian intellectuals as prominent as Italo Calvino, Umberto Eco, Franco Fornari, and Leonardo Sciascia, was badly misunderstood:[33] "I have never talked about life in general, but I have always talked about *this* life, *this* mother, *this* womb, *this* unborn baby."[34]

What should interest us the most here is such insistence on the singularity and concreteness of life. Pasolini seems here unable to fully realise that, outlining the way in which a *singular* life is sacred, he also concomitantly promotes an alternative idea of life as a whole. He is no doubt suggesting that *each* new birth constitutes *a* life as absolutely singular: "*This* single, concrete human life," he says.[35] However, as he himself appears to imply without fully acknowledging it, what is at stake in such an irreducibly singular life that can be counted as one is human life conceived as a *totality*, or better, its survival. Singular one-lives – first and foremost that which, having been born, will trigger an irreversibly lethal Malthusian crisis – must be understood against the background of life in general as a differential One, a One of difference. Perhaps we should invoke here, in passing, the capitalised notion of "a life" (*une vie*) advanced by Deleuze in the last essay he wrote before his death: "A LIFE [...] the immanence of immanence, absolute immanence: it is complete *power*, complete *bliss*."[36]

2 birth as affirmative biopolitics and the de-generation of life

The work of Roberto Esposito develops an original philosophical approach to the question of biopolitics that has only recently started to receive the international attention it deserves. Esposito believes that, despite the popularity it has acquired in many academic quarters, biopolitics is still lacking any kind of adequate categorisation. First and foremost, this would be revealed by the fact that it can be interpreted in almost diametrically opposite ways. Esposito thus starts off from the premise that we should avoid embracing both Agamben's "negative" assessment – which reduces biopolitics to "an antinomic repetition of sovereign power's lethal paradigm" – and Negri's "euphoric" reading – which identifies it with "a power of life that is always excessive and finally subversive."[37] More importantly, he suggests that the very possibility of such radically different definitions is to be related to the structure of the notion of biopolitics as it was originally formulated by Foucault. For Esposito, Foucault never really provided a definitive answer to the crucial issue of the link between modern sovereignty and contemporary biopolitics. Inasmuch as both regimes seem to acquire their ultimate meaning only against the background of death – the thanatopolitical outcomes of twentieth-century biopolitics do not differ on this point from the sovereign's right to kill – it remains uncertain whether one regime actually excludes the other.

> On the one hand, [Foucault] conjectures something like the return of the sovereign paradigm within the biopolitical horizon. On the other hand, and simultaneously, he advances the opposite hypothesis: [...] it was precisely the final disappearance of the sovereign order that unleashed a vital force that was so full that it overflowed and turned against itself.[38]

According to Esposito, this impasse in periodisation is the consequence of a much more serious problem in Foucault's analysis of biopolitics: the French philosopher would only think the connection between politics and life in an external way. Although he grasps their reciprocal implication, they nevertheless remain "two poles [...] closed within separate orbits [...] themselves undefined in their outline and quality."[39] In response to such perceived vagueness, Esposito intends to investigate the organic nexus between politics and life by means of the notion of "immunisation":

> Rather than being superimposed or juxtaposed in an external form that subjects one to the domination of the other, in the immunitarian paradigm, *bíos* and *nomos*, life and politics, emerge as the two constituent elements of a single, indivisible whole that assumes meaning only from their relation. Immunity is not simply the relation that joins life to power, but life's power of preservation [*il potere di conservazione della vita*]. Contrary to what is presupposed in the concept of biopolitics – understood as the result of an encounter that arises *at a certain moment* between the two components – from this standpoint, there is no power external to life, just as life is never given outside of relations of power. From this perspective, politics is nothing other than the possibility, or the instrument, for keeping life alive.[40]

These specifications allow Esposito to define immunisation in terms of "*negative protection*" (*protezione negativa*).[41] In other words, given what we have just seen, from an immunitarian perspective, power does not simply negate life nor does it simply protect and reproduce it. Immunitarian power is rather that by means of which life, in negating itself, also inherently preserves and reproduces itself.

Having said this, Esposito also emphasises the fact that, as is shown by auto-immune diseases, an excess of immunisation runs the risk of annihilating life. For him, today's global politics is ruled by an "immunitarian obsession" – epitomised by the pre-emptive War on Terror – that, having expanded immunisation beyond a critical threshold, has itself become a danger for life. In opposition to such "politics *on* life" – a biopower that assumes life as its separate subjected object and thus paradoxically contradicts the very basis of the immunitarian paradigm – philosophers need to formulate a "politics *of* life," or "life's politics."[42] Surprisingly enough, Esposito claims that such an "affirmative biopolitics," a "new global political order founded upon the difference of the elements that compose it," is already outlined metaphorically in the notion of biological birth.[43] More specifically, the biological process of gestation and birth would as such challenge and refute the validity of any notion of immunisation based on sheer self-preservation: indeed, not only does the mother's immune system tolerate the presence of the foetus's different immune system inside her body, but even "preserves it from the threat of abortion in a way that is directly proportional to its genetic diversity."[44] In other words, immunity does not correspond in this case to a barrier, let alone a weapon, against what is different, but rather a "filter" by means of which one communicates with it. Birth constitutes the ground zero of immunisation since, in these circumstances, the protection of life exercised by the mother over her own body and the foetus perfectly coincides with the pure affirmation of the latter's life. Here, conservation meets innovation and *immunitas* becomes indistinguishable from *communitas*. A community as "collective existence" and "social circulation" is thus in the first instance always the indirect result of a maternal immunity that donates life in the form of a gift – a *munus* – to a "singular but also infinitely plural" newborn.[45]

While Agamben only sketches a pessimistic genealogy of the modern link between birth and nation – one that would inevitably culminate in their paradoxical disjunction in the space of the extermination camp[46] – here, Esposito follows Pasolini's lesson and locates birth as a potentially emancipatory notion at the centre of his analysis of biopolitics.[47] But how, more precisely, should we compare and contrast Esposito's and Pasolini's considerations on the biopolitical status of birth? Most importantly, does Esposito's detailed philosophical investigation of biopolitics in general clarify, if not overcome, Pasolini's provocative polemical stance?

On one level, initially, we could suggest that Esposito helps us to unmask the thanatopolitical implications of Pasolini's own radical critique of biopolitics in so far as he resolutely condemns any discourse on the alleged degeneration of life. Analysing Nietzsche's late works, Esposito shows us how an affirmative discourse on generation can easily turn into a thanatopolitical discourse on de-generation. If, for Nietzsche, life as such is inherently a will to power that can only affirm itself by means of a continual expansion, and if this expansion ultimately amounts to a constitutive "self-dissolving drift,"[48] his thought is inevitably based on a structural antinomy. As Esposito writes:

> [According to] the indissolubly tragic nature of the Nietzschean perspective [. . .] the survival of a force contrasts with the project of its strengthening [*potenziamento*]. Limiting itself to survival, it weakens itself, [. . .] *degenerates*, which is to say moves in the direction opposite to its own generation. But, on the other hand, if this is the case, must we draw the paradoxical conclusion that to expand vitally, an organism has to cease to survive? Or, at least, that it must defy death?[49]

Esposito draws two major conclusions from these remarks. Firstly, for Nietzsche, any immunitarian endeavour to protect life against its expansive involution only achieves a privileging of weak life over strong life and, consequently, an acceleration of the process of degeneration. Secondly, Nietzsche's own negation of immunisation unintentionally falls back into a negative immunitarian logic, that is, it does not challenge death. This happens as soon as unprotected strong life is enjoined to protect itself from the contagion of weak protected life. At this stage, Nietzsche's will to power, originally thought in terms of unrestrainable excess and abundance, is contradictorily reduced to a hyperimmunitarian equation for which the expansion of strong life is directly proportional to the compression or even the elimination of weak life. For Esposito, Nietzsche's undeniable racism – which is both intra- and inter-ethnic – can thus be understood only from a thanatopolitical perspective: in so far as life entails "continually shedding something that wants to die," the life of some people must be completely opposed to – and affirmed against – the non-life of some other people.[50]

Following on from this, we might suggest that Pasolini's treatment of sacred life is vulnerable to a similar criticism. As we have seen, he believes that the hedonistic consumerism and sexual promiscuity imposed by neo-capitalist power necessarily entails the destruction of popular culture. But besides promoting new homogenising psychological traits through the simplification of expressive codes – verbal and non-verbal – such "anthropological genocide" is also concomitant with a *corporal* degeneration that has allowed the emergence of nothing less than a new human race.[51] Pasolini thus openly speaks of a "*biological* deterioration" from which "the man of the mutation" emerges, a "mostrous creature of the system," who is deprived of any power and yet, at the same time, is aggressively "ready to do anything" for the sake of consumerism.[52] For a while, during the years of the post-Second World War economic boom, capitalism deluded us into believing that the conquests of medicine and an improved diet were paving the way for a "better" human race. But this was only a brief illusion. Just as, in an explicit attack on Darwin, Nietzsche observes that the current "selection" of life entails "the cancellation of happy cases, the uselessness of types that are highly successful, the inevitable victory of the average and even of those *below* average,"[53] so Pasolini claims that "the new generation is infinitely weaker, uglier, paler and sicker than all other generations one can remember."[54] In his unfinished and posthumously published novel *Petrolio*, Pasolini goes as far as advocating that the genetic degeneration of the new generation can also be objectified into specific physical deformities such as obesity, bandy legs, being excessively short or lean, having flattened or misshapen noses, having huge mouths and prominent teeth, and so on.[55]

More significantly, scrutinising Pasolini's last writings, it is possible to recover a progressive shift from an ecological critique of biopolitics to an attempt to prevail over its potentially disastrous consequences that acquires dubious thanatopolitical connotations. While the articles on abortion collected in *Scritti corsari* – written mainly in January and February 1975 – insist on the reversibility inherent to the sacred character of life following a Malthusian demographic argument, the later texts included in *Lettere luterane* – written mainly in the spring and summer of 1975 – adopt a disturbingly eugenic tone. In other words, in the few months before being murdered, Pasolini seems to be increasingly interested in identifying *which* sort of sacred life should be regarded as damned.[56] If, in *Scritti corsari*, all new lives are as such considered to be damned due to overpopulation and the subsequent possibility of the extinction of the human species, in *Lettere luterane* Pasolini clearly qualifies supernumerary births and thus neatly separates them from sacred life as blessed birth. In doing so, he considerably reduces the distance between his fight against biopolitics and the "incessant decision on value and nonvalue" – that is, on what, within *sacertas*, distinguishes blessedness from damnation – which Agamben has good reasons to associate with Nazi eugenics.[57]

How, more specifically, does Pasolini characterise sacred life and thus inadvertently compromise his courageous denunciation of hedonistic consumerism? In an article dated 22 May 1975, and tellingly entitled "They live, but they should be dead," he seems to understand damned birth in two different, albeit overlapping, ways. On the one hand, he speaks of "those who were destined to be dead," that is, children who in the past would not have survived without the help – the immunitarian protection – of medical technology. On the other hand, he speaks of those who are not loved. Although today no child is ever as blessed and loved as he would have been in the past due to the risk of a demographic catastrophe, it is nevertheless the case that "those who were even 'torn' from the innocent death of infancy feel with still more violence their guilt for being alive."[58] Here, not only is Pasolini after all implicitly insinuating that the need for technologically assisted protection is in a child inversely proportionate to his being loved since the moment of conception, but he is also relating unloved life that would not have survived in the past to excessive births. Quite crudely, the births of unloved children who could not survive without the "artificial" support of medicine are those that will set off an irremediable Malthusian crisis.

Having said this, in a passage from the same article, Pasolini seems to acknowledge the importance of *not isolating and personifying* sacred life when he specifies that, in today's ecological predicament for which birth can never be welcome, "the judgement on what is blessed and what is damned must be suspended." However, in the following sentence, he also blatantly contradicts himself by adding that "those who are born 'in excess' [*coloro che nascono "in più"*] are definitely damned."[59] Here, Pasolini's reflection on biopolitics reaches its point of maximum tension. It is indeed precisely the replacement of the generic indecidability concerning the sacredness of life with the identification of a well-defined surplus of humanity that transforms his attack on biopolitics into a potential source of inspiration for an extensive thanatopolitics. From this perspective, the very legalisation of abortion on specific ecological "extenuating circumstances" – as opposed to its legalisation as such – casts a sinister shadow on Pasolini's suggestions. The foetus to be aborted – or the child that should be left to die – is in this case *determined* as the supposedly more unloved child who, *thanks to* technological immunisation, will eventually cause the extinction of the human species. Beyond (if not against) Pasolini's intentions, we could argue that such a child thus comes to occupy the universally singularisable position of a type of generalised *homo sacer* unexplored by Agamben, one that could easily be appropriated by naïve pro-life rhetoricians: each one of us as a virtually unloved child could have been "killed" with impunity by those who fix politically the demographic threshold beyond which birth is deemed superfluous, or even detrimental, to the perpetuation of the species.[60]

3 one divides into two-as-one: a transcendent notion of birth

Esposito's convincing analysis of the "logical and semantic chain" that structurally relates any discourse on the degeneration of life to a discourse on regeneration and ultimately genocide – "regeneration overcomes degeneration through genocide"[61] – should make us wary of any decision to determine sacred life. Yet, in a number of passages from *Bíos* (2004) – which I would regard as his most important work to date – Esposito is himself inclined to suspend the undecidable reversibility of sacredness in an affirmative way: in these crucial instances, birth as the site of bare life is determined as a priori blessed.[62] Esposito does not deny that birth may be appropriated as a negative biopolitical category liable to justify the worst thanatopolitical crimes: the Nazis' pro-natility campaign that intended to regenerate the German people by means of official eugenic protocols was, for example, inextricable from the promulgation of laws that imposed sterilisation on three hundred thousand people. Having said this, it is also clearly the case that, for Esposito, the event of birth as such – in the last instance even the "forced birth" that follows ethnic rape – eventually always entails that "the force of life prevails once again over that of death."[63] While a statement like this runs the risk of echoing the most stereotypical and reactionary Christian pro-life arguments – especially when corroborated by the alleged evidence that "Rwandan war mothers [. . .] *all* declared their love for their children born from hate"[64] – it also shows why Esposito interestingly thinks of Nazism as a thanatopolitics which, first and foremost, dreaded birth as a "spontaneous and unexpected" biological phenomenon.[65] If what Esposito names "anticipatory suppression of life" became one of the most consolidated immunitarian apparatuses of the Third Reich, this happened because ungoverned birth always threatens to unmask the "original duplicity of the Origin" (*originaria doppiezza dell'Origine*),[66] which as such jeopardises any identitarian politics:

> Rather than enclosing the extraneousness within one and the same body [. . .] and so obliterating it, birth ejects what is within the maternal womb into the external world. It does not incorporate, but excorporates, exteriorises, turns inside out. It does not presuppose, or impose, but exposes someone (male or female)

> to the event of existence. Therefore, it cannot be used, in either a real or metaphorical sense, as a protective apparatus of the self-preservation of life.[67]

It would therefore be misleading to assume that the Nazi notion of the "nation" was directly rooted in the natural birth of the citizens of German blood. For Esposito, it was not birth that decided the political status of the inhabitant of the Third Reich but rather his "position in the political-racial calculation that *predetermined* the value of his birth."[68]

At this stage, how should we interpret the fact that Esposito seems to think the original duplicity of birth together with its a priori blessedness? I believe that it is on this level that his comprehensive examination of biopolitics clarifies the vitalistic horizon of Pasolini's "ecology" but does not yet manage to overcome it. In a few words, both authors start off from the very same premise: before and outside of any symbolic (political and legal) constraining imposition *on* life,[69] life *is* in itself *a* life as a *unity* of difference that is as such sacredly blessed. In spite of conceiving the origin as an "original in/originarity" (*in/originarietà originaria*),[70] Esposito does not question the assumption according to which birth is, at the same time, contradictorily considered as a "subdivi[sion of] the one – the body of the mother – into two."[71] On the one hand, we are confronted here with life as an "irreducible difference," an in-original generative two that always already immanently contains in itself all successive births as a multiplying "plurality of infinite numbers."[72] On the other hand, this very difference ultimately amounts to nothing less than an original transcendent One given that birth is plainly understood as "a one – the body of the mother – that makes itself into two, offering a part of itself, of its life, to another life."[73] The "biological" life of the mother as a (differential) One that can divide itself into parts – again, "the original duplicity of *the* Origin" – is problematically taken for granted. In other words, if birth as origin is double, why does Esposito think what "precedes" the origin, "the body of the mother," as a (differential) One? Isn't he in this way unintentionally supporting Pasolini's quite simplistic idea that "the maternal womb [hosted] an existence in which reigned the plenitude of a paradise that has definitively been lost"?[74] And most decisively, to remain within an immanentist perspective, shouldn't Esposito rather assume that the one of the *mother's* body emerges only retroactively as a consequence of the – literally material – coming to light of the "original duplicity of the Origin" (birth) out of a "primordial" not-one (foetal gestation)?[75]

Esposito's intention to investigate biopolitics, beyond Foucault, as an immanent nexus in which and for which life and politics are – and have always been – inextricable from one another is thus compromised by his own transcendent stance on life. As we have seen, he believes that life never gives itself outside of relations of power. Life is for him "touched, crossed and modified by history" and consequently not an "unalterable given" that can be conceptualised scientifically as something static.[76] Yet, at the same time, Esposito has not fully disposed of the traditionally substantialist idea of life as a "presupposition" (*presupposto*): he simply replaces it with the vitalistic idea of differential life as a "place" (*posto*) of blind "forces," "natural impulses," "raw matter," a "primordial intensity" that, after all, inherently aims at expanding itself independently of politics.[77] The very use of such terminology is clearly symptomatic of an implicit primacy of life as power – or, more precisely, as will to power – over political power as life.[78] This impasse in Esposito's thought becomes obvious when, in formulating his notion of immunisation as negative protection, he thinks of life against the background of negation. At first sight, his claim that "negation is not a form of violent subjection [*assoggettamento*] that power imposes on life from the outside, but rather the inherently antinomic way in which life preserves itself through power" appears to be consistent with an immanentist logic.[79] However, on closer inspection, we soon realise that what remains unthought in such a paradigm of immunisation is the reason why life as "power [*potenza*] to expand"[80] is assumed to precede – logically if not chronologically – power, or better politics, as protective negation.

Given the immanent nexus for which "life and politics represent the two constituent elements of a single, indissoluble whole that acquires meaning only on the basis of their relation" shouldn't life rather be thought of as inseparable from negation? Or, from a slightly different perspective: if life antinomically needs to preserve itself through politics in a negative way, how can it in the first place be defined in terms of sheer affirmative expansion? If the power of life, detached from politics, would "naturally" cause the destruction of life, isn't it more appropriate to understand "life" without politics – and, more to the point, human "life" subtracted from symbolic negation – in terms of *im-potence*?

In the concluding section of *Bíos*, Esposito acknowledges his philosophical indebtedness to Deleuze and focuses on his last work, "Immanence: A Life." It is in this context that Esposito's problematic notion of life and attempt to formulate an affirmative biopolitics of immanence show a surprising proximity to Pasolini's understanding of sacred life as singular blessed life. As we have already seen, according to Deleuze, pure immanence is "A LIFE, and nothing else [. . .] it is complete *power*, complete *bliss*."[81] Resuming implicitly Pasolini's insistence on the sacredness of "*this* single, concrete human life," Esposito rightly emphasises Deleuze's stress on "a life" as "a life like this," the "like this" of a life. Such "a life like this" is, in the first place, singular precisely in that it exceeds individuality and resolves itself into an impersonal given. The "like this" of a life is not individualisable but generic since it pertains to a genus, mankind as such. Yet, at the same time, each "life like this" is also utterly singular in the sense that it unmistakably differs from other singular lives. Such "impersonal singularity" is, for Esposito, significantly enough, epitomised by the newborn baby, who is "similar to all the others, but different from each of them by the tonality of the voice, the intensity of a smile, the sparkle of a tear."[82]

Summarising his views, Esposito identifies the basic presupposition of any anti-thanatopolitical "biophilosophy" of immanence in the substitution of the definite notion of "the" life with the indefinite notion of "a" life. Once again, I think we should challenge his argument. Inasmuch as "*a*" life remains after all, in spite of its impersonal singularity, *one* life, a single unified life of pure difference,[83] shouldn't we rather locate real immanence on the level of life that is not "a" life? This would obviously not amount to non-life (a denial of life) but to life that is not-one (to life as conceivable solely within the – symbolic – horizon of negation). Most importantly, it is only at this stage that we would be advancing a notion of immanence freed from any dependence on Christian sacrality. The impersonal singularity of the newborn baby – the "*homo tantum*" who, according to Deleuze, "attains a sort of beatitude"[84] – still relies on a transcendent Christian notion of birth as the passage from the One to the two-as-One. Such a notion – to which, as we have seen, Esposito ultimately subscribes – must be overcome with a materialist notion of birth as the *immanent* passage from the not-one of pre-symbolic life to the two as the truly inextricable nexus between human life and politics.

4 birth as legal obligation: esposito with john paul ii

In the first two chapters of the encyclical letter *Evangelium Vitae* (1996), Pope John Paul II spells out for Christians and "all people of good will" the basic tenets of the Church's stance on human life. We could abridge them as follows. Firstly, life is a sacred reality that God entrusts to us and is consequently inviolable (2, 22); inasmuch as it is a manifestation of God in the world, life is always good (34). Secondly, life is total and full (31); life was always in God and God must be understood as "the life of the 'Eternal One'" (37). Thirdly, man – his life – was born of God (37, 39). This is valid for all men, not only Adam: God makes himself present at every birth in that, "when a new person is born [. . .] he brings with him into the world a particular image and likeness of God himself" (43). Joining these three fundamental principles together, it is possible to infer that the sacredness of human life has its foundation in the fact that it is born of God as the life of the eternal One (37, 39). Most importantly, given the appearance of sin in history – a rebellion of man against his Creator through which death came into the world (36, 7)[85] – man is redeemed and participates in the eternal life of God only if he hears the words of Jesus Christ. These words reveal nothing other than the truth of "the fullness of life" in which men must have faith: "This is eternal life, that they may know you the only true God [. . .] the Holy one of God [. . .] and Jesus Christ whom you have sent" (Jn 17: 3, 6: 68–69). As a consequence, the Gospel – the "good news" that Jesus preached and *incarnated* – is by definition a Gospel of Life, an *Evangelium Vitae*, that first and foremost announces the Birth of the Christ Child as the Saviour (1). The new law of the Gospel is a "law of life" (48) – as Agamben has remarked, Christianity challenges the traditional form of the law of sovereign power with the form of life of Christ qua the Gospel[86] – which as such "fully protects human life" (48).

At this stage, we might be tempted to argue that the Pope's emphasis on the protection and defence of life – which also invariably informs everyday pro-life parlance – is precisely what Esposito's critique of the thanatopolitical logic of immunisation wishes to distantiate itself from. However, we should not lose sight of the fact that, for Esposito himself, birth represents a special kind of immunisation thanks to which the protection of life for once perfectly coincides with its pure affirmation. Furthermore, as we have seen, birth as an immunitarian gift is, for Esposito, inextricable from the communitarian social circulation it founds and sustains. In his encyclical letter, John Paul II makes a similar argument: life is a gift of God and "every human community [as] political community" can preserve itself only if the sacred value of human birth is being recognised and respected as a primary right (2). Both the Pope and Esposito therefore understand the gift of *biological* life in terms of a *legal obligation* that, as such, directly constitutes the fabric of humanity *tout court.* More specifically, according to John Paul II:

> By accepting God's gift, man is obliged to maintain life in this truth [the truth of "the fullness of life"] which is essential to it. To detach oneself from this truth is to condemn oneself to meaninglessness and unhappiness, and possibly to become a threat to the existence of others, since the barriers guaranteeing respect for life, in every circumstance, have been broken down. (48)

As for Esposito:

> [Birth as] donative exposure involves an obligation for whoever receives it. Hence, [this is valid] for all men, every man in relation to every other man. Such obligation should not be understood in an ethical sense, or [at least] only in an ethical sense. Before this, and more profoundly, [it should be understood] in an ontological sense. That is to say that such obligation precedes ethics since it precedes the very dimension of *humanitas* on which all ethics depends. It precedes it because it is rooted in the matrix where *humanitas* is not yet such [. . .] bare life, not in the sense of a brute biological layer, but in that of exhibited life, life without any protection other than the force of life itself.[87]

Here, Esposito follows very closely the "law of life" of the Gospel that, as we have seen, "fully protects human life." In the end, he is suggesting that the acknowledgement of biological birth as a binding gift is by itself a necessary and sufficient condition for the foundation of the *humanitas* of man, his symbolic – and ultimately blessed – communitarian dimension. And this is so because, before becoming the object of any bio-ethical legislation, birth and the ontological obligation it carries with it directly issue from "the force of life itself," the ur-protection of a life that is otherwise as yet unprotected.

More generally, Esposito appears to develop an ambiguous and, in my opinion, inconsistent relationship with the Christian idea of life. On the one hand, his notion of birth cannot be separated from his attempt to think impersonal singularity: the latter is obviously in open conflict with the importance that the Church bestows on the protection of life as the integrity of the individual *person* (40).[88] On the other hand, inasmuch as Esposito conceives birth as "the actual place in which *a* life makes itself two,"[89] such a "blessed" *One* of difference, which we discussed above, presupposes a stance on the differentiality of life which ultimately resembles that of John Paul II's *Evangelium Vitae*, and is perhaps even identical with it. According to the encyclical letter, God as the One of eternal life creates human life as a different life which nevertheless initially participates fully in divine life. Although this is not made explicit by the Pope, the birth of man can thus be regarded as the passage from the One to the two-as-One: divine and eternal life as One becomes One (God) and two (God as different from his human image) in One. Although man's fall decreed the emergence of "life in time," thanks to Jesus Christ's redemptive sacrifice, eternal life continues to be "the life of God himself and at the same time the life of the children of God" (38). Conversely, "life in time [. . .] is the fundamental condition, the initial stage and an integral part of the entire unified process of human existence" (2) since "God's eternal life is in fact the end to which our living in this world is directed and called" (30).

To conclude, I would claim that Esposito limits himself to renaming the Christian One as the difference between transcendence and immanence – which is itself contained within transcendence – in terms of the putatively immanent One of difference, which he derives from his interpretation of Deleuze. Esposito antinomically thinks the duplicity of the in-original origin together with the precedence of the One over the origin and finally subjects the former to the latter. My contention is reinforced by his own attempt to reformulate, in what he deems to be a "non-Christian" way, a notion that is theologically contiguous with that of birth, namely, incarnation.[90] Unlike incorporation, that "tends to unify a plurality, or at least a duality," incarnation would "separate and multiply in two *what was originally one* [*ciò che all'origine era uno*]."[91] Here, it is evident that not only is the duplicity of the in-original origin contradictorily supplemented by the pre-existence of the One but it is *itself* straightforwardly conceived as a *unity*. In other words, the supposedly immanent passage from the one to the two ("the-two-in-one" or "the-one-that-makes-itself-two") as opposed to the transcendent passage from the two to the one is then nothing other than a variation internal to the same One, what Esposito himself names – precisely in his discussion of "the-one-that-makes-itself-two" – "an *initial* identity."[92]

notes

1 P.P. Pasolini, *Saggi sulla politica e sulla società* (henceforth SPS) (Milan: Mondadori, 1999) 372.

2 See also Italo Calvino's critical considerations in "Che cosa vuol dire rispettare la vita," *Corriere della Sera* 8 Feb. 1975.

3 SPS 372.

4 Ibid. 376, 384.

5 Ibid. 376–77.

6 Ibid. 376.

7 Ibid. 384.

8 Ibid.

9 Ibid. 459, 1547. On Pasolini's attempt to develop a universalist Marxian politics of poverty as a politics of charity (*agape*), see L. Chiesa, "Pasolini, Badiou, Žižek und das Erbe der christlichen Liebe" in *Wieder Religion? Christentum im zeitgenössischen kritischen Denken*, eds. M. De Kesel and D. Hoens (Vienna: Turia, 2006) 107–26.

10 SPS 321.

11 A. Negri, "The Italian Difference" in *The Italian Difference. Between Nihilism and Biopolitics*, eds. L. Chiesa and A. Toscano (Melbourne: Re.press, 2009) 16–17.

12 However, drawing on an official publication of the National Socialist regime from 1942, Agamben does not fail to observe that the Third Reich's medical politics ultimately aimed at a "logical synthesis of biology and economy" (*Homo Sacer. Sovereign Power and Bare Life* (Stanford: Stanford UP, 1998) 145).

13 SPS 402. On the perfect overlapping of the "society of producers" with the "society of consumers" in today's bio-economy, see L. Bazzicalupo, *Il governo delle vite. Biopolitica ed economia* (Rome and Bari: Laterza, 2006) 118–25. See also Andrea Fumagalli's twelfth thesis on

contemporary capitalism in his contribution to the present issue. On the monetary dimension of *sacertas* in Roman law, see G. Agamben, *The Sacrament of Language: An Archaeology of The Oath* (Stanford: Stanford UP, 2011) 64–65.

14 SPS 652.

15 Ibid. 263.

16 See, for instance, SPS 1546–47. See also P.P. Pasolini, *Saggi sulla letteratura e sull'arte – tomo secondo* (Milan: Mondadori, 1999) 2860. On how to interpret in this light the legalisation of divorce in Italy, see SPS 516.

17 "The new mode of production [...] *is not just a production of commodities, but of humanity* – as a matter of fact, this is what the elementary law of political economy teaches us" (SPS 657).

18 SPS 1537. For a further discussion of these issues, see L. Chiesa, "Pasolini and the Ugliness of Bodies" in *In Corpore: Bodies in Post-Unification Italy*, eds. L. Polezzi and Ch. Ross (Madison: Farleigh Dickinson UP, 2007) 208–27.

19 Such simplifications are indirectly promoted by Agamben himself whenever he proposes facile analogies or antonymic oppositions between, for example, *homo sacer* and the detainees in Guantánamo; the prevailing governmental paradigm and the figure of the tourist; the nudity of those who are being baptised and that of sunbathers; the hairy body of the resurrected and the depilated body of fashion models or pornstars...

20 SPS 375.

21 Ibid. 1548.

22 Ibid. 373.

23 Ibid. 397.

24 Ibid. 1547–48. This complex reasoning should closely be confronted with Agamben's subtle interpretation of the *katéchōn*, the ultimately self-destructive constituted power that endeavours to repress the *katárgēsis*, a state of anomie which suspends the law from within (see, for instance, G. Agamben, *The Time that Remains* (Stanford: Stanford UP, 2005) 110–12). In *The Kingdom and the Glory*, Agamben briefly acknowledges his proximity to Pasolini's reflection on political illegality when he states that the remark of one of the fascist sadists in *Salò*, according to which "the only real anarchy is that of power," is "perfectly serious" (G. Agamben, *The Kingdom and the Glory* (Stanford: Stanford UP, 2011) 64).

25 SPS 378.

26 Ibid. 291.

27 Ibid. 711–12, 672.

28 On how a "dis-alienation of love" by means of a "universal pedagogy of love" that also includes sexual education and contraception should be regarded as the right alternative to the legalisation of abortion, see ibid. 1547, 1549.

29 Ibid. 697, 712.

30 Ibid. 1480, 1474.

31 Ibid. 386–87.

32 Ibid. 543.

33 The names of the writers Natalia Ginzburg, Giorgio Manganelli, Dacia Maraini and Alberto Moravia should also be added to this impressive list. The depth and intensity of the debate, published without delay in some of the most important national newspapers – such as *Corriere della Sera*, *La Stampa*, and *Il Manifesto* – would deserve an article in itself.

34 SPS 401.

35 Ibid. 384.

36 G. Deleuze, "Immanence: A Life" in *Pure Immanence. Essays on a Life* (New York: Zone, 2001) 27; my emphasis.

37 R. Esposito, "Prefazione" in Bazzicalupo, *Il governo delle vite* vii; R. Esposito, "Biopolitica, immunità, comunità" in *Biopolitica. Storia e attualità di un concetto*, ed. A. Cutro (Verona: Ombre Corte, 2005) 158. On how Agamben's thought could also be considered as an *affirmative* bio-theo-politics, see L. Chiesa, "Giorgio Agamben's Franciscan Ontology" in *The Italian Difference* 149–63.

38 "Biopolitica, immunità, comunità" 159.

39 Ibid. 160.

40 R. Esposito, *Bíos: Biopolitics and Philosophy* (Minneapolis: U of Minnesota P, 2008) 45–46; translation modified.

41 Ibid.; translation modified.

42 "Biopolitica, immunità, comunità" 164.

43 Ibid. 165; my emphasis.

44 Ibid.

45 Ibid. 166, 161.

46

> The growing dissociation of birth (bare life) and the nation-state is the new fact of politics in our day, and what we call *camp* is this disjunction; The camp [. . .] is produced at the point at which the political system of the modern nation-state, which was founded on the functional nexus between a determinate localization (land) and a determinate order (the State) and mediated by automatic rules for the inscription of life (birth or the nation), enters into a lasting crisis, and the State decides to assume directly the care of the nation's biological life as one of its proper tasks. (*Homo Sacer* 174–75)

47 In a short chapter of his most recent book, *Pensiero vivente. Origine e attualità della filosofia italiana* (Turin: Einaudi, 2010) 192–206), Esposito acknowledges the importance of Pasolini's thought for contemporary biopolitical speculation. However, he does not deal with his treatment of birth, or question the many similarities in their analysis of this issue.

48 *Bíos* 81, 88; translation modified.

49 Ibid. 94; translation modified.

50 Ibid. 98. For Esposito, there is nonetheless also another Nietzsche, whose philosophy is anti-thanatopolitical and defies death (see ibid. 101–09).

51 See, for instance, SPS 675.

52 Ibid. 1449, 1527; my emphasis.

53 *Bíos* 95.

54 SPS 589. For a serious biological account of the "dangers of genetic degradation in modern societies" from a perspective that is both Darwinian and politically compatible with what the author, quarrelling with Lysenko, calls "a socialist humanism that is really *scientific*," see the last chapter of J. Monod, *Chance and Necessity: An Essay on the Natural Philosophy of Modern Biology* (London: Collins, 1972).

55 P.P. Pasolini, *Petrolio* (Turin: Einaudi, 1992) 333. I have discussed this in detail in "Pasolini and the Ugliness of Bodies."

56 The same could be said about Pasolini's last film, *Salò*, which was also shot in the spring of 1975.

57 *Homo Sacer* 153. In a letter to Moravia, published by *Corriere della Sera* on 30 Jan. 1975, in which he replies to the accusation of defending abortion from a position that is both Catholic and sexophobic, Pasolini refers in passing to Himmler: like the latter, the vast majority of Italians would regard Pasolini's own life as "unworthy of being lived" due to his openly professed homosexuality (SPS 380). On how neo-capitalism's hedonistic regimentation is seen by Pasolini as a new form of Nazism, see my "Pasolini and the Ugliness of Bodies."

58 SPS 589.

59 Ibid. 588.

60 In a 1975 interview, Pasolini does not hesitate to define abortion as a "legalisation of homicide" (ibid. 1545). It must, however, be stressed that, in accepting it on the basis of extenuating ecological circumstances, he does after all adopt a position that is irreconcilable with Catholic pro-life orthodoxy. In this light, he affirms that he "does not support either side" of the debate, especially because "those who are against [abortion] do not really oppose it. They are themselves part of the same prevarication as long as they omit to pose preliminarily the question of coitus" (ibid. 1547). And yet, shortly after, Pasolini seems to fall back into pro-life dogmatism when the Malthusian premises of his own argument about *Lebensraum* are put into doubt: "Have these previsions trusted by the de-populators ever established with certainty that our planet is an insufficient vital space?" (ibid. 1548–49).

61 *Bíos* 137.

62 This is ultimately also Pasolini's stance: supernumerary births are damned only in so far as

they compromise the perpetuation of the human species as originally blessed life.

63 *Bíos* 7.

64 Ibid.; my emphasis, translation modified.

65 Ibid. 145; translation modified.

66 Ibid. 176; translation modified.

67 Ibid.; translation modified.

68 Ibid. 145; my emphasis.

69 Pasolini thus writes: "Culture gradually impoverishes and simplifies nature. The more one lives in the state of nature, the more the code is complex, alive" (SPS 1461).

70 *Bíos* 175; translation modified.

71 Ibid. 176.

72 Ibid.

73 "Biopolitica, immunità, comunità" 166.

74 SPS 1545.

75 Such "primordial" not-one – which also remains *simultaneous* with duplicity after the latter's emergence – corresponds to a concrete instantiation of what I have defined elsewhere as the "undead." For an in-depth analysis of this concept, see L. Chiesa, "Notes Towards a Manifesto for Metacritical Realism" in *Beyond Potentialities? Politics between the Possible and the Impossible*, eds. M. Potocnik, F. Ruda, and J. Völker (Berlin: Diaphanes, 2011).

76 *Bíos* 29–30. Esposito's repeated call for a return to "Darwin's original intuition" is thus always rightly accompanied by a stringent critique of naturalistic reductionism. See his contribution to this issue.

77 Ibid. 30, 48, 59.

78 The same applies to *Pensiero vivente*, Esposito's otherwise excellent genealogy of the interrelation between life, politics, and history in modern Italian thought. When, for instance, he appropriates to his own philosophical project Machiavelli's, Bruno's, and Vico's "common thrust towards an origin [...] full of vital energy" (ibid. 85), the basic "natural energy present in each man" (Machiavelli), an "inexhaustible Substance" (Bruno), "a living body" (Vico) (ibid. 52, 62, 77), he does so without questioning the vitalistic presuppositions taken for granted by these authors.

79 Ibid. 46; translation modified.

80 Ibid.

81 "Immanence: A Life" 27.

82 *Bíos* 193. Here, Esposito follows Deleuze's arguments and examples closely. Esposito has further investigated the notion of the impersonal in relation to affirmative biopolitics in *Terza persona. Politica della vita e filosofia dell'impersonale* (Turin: Einaudi, 2007).

83 The very last page of *Pensiero vivente* provides strong evidence to support my reading:

> Life should neither be postponed nor presupposed by the subjects who incarnate it; rather, it should be thought as the *living substance* of their infinite singularity [...] However, [life] is only *a single life*, no matter how many times it reproduces; it is always the same and always different in each of us. (Ibid. 265; my emphasis)

84 "Immanence: A Life" 29.

85 God is life without death ("God did not make death") and death concretely entered the world only by means of a killing, "the killing of Abel by his brother, Cain." The latter amounts to a violation of the "kinship of flesh and blood" which is today perpetuated "within the relationship between parents and children [...] in abortion" (7). It is worth dwelling briefly on the figure of Cain from the standpoint of *sacertas*. His being the first killer not only introduces death – which is thus *not* a direct consequence of the sin of our first parents ... – but, following God's punishment, also renders him the first "fugitive and wanderer on earth," one who, however, *suspends the very suspension of law* embodied by *homo sacer*. Replying to Cain's fear – "whoever finds me will slay me" – the Lord says "if any one slays Cain, vengeance shall be taken on him sevenfold!" and puts a mark on him "lest any who came upon him should kill him."

86 G. Agamben, *The Time that Remains: A Commentary on the Letter to the Romans* (Stanford: Stanford UP, 2005) 27.

87 "Biopolitica, immunità, comunità" 166.

88 Esposito gives a convincing account of the Roman and Christian origins of the notion of the person in *Terza persona*. According to him, such a notion should rather be considered as an apparatus whose primary effect is the separation of rationality and the will from the biological animality of man.

89 *Bíos* 108; my emphasis.

90 Ibid. 168.

91 Ibid. 167; my emphasis.

92 Ibid.

giorgio agamben

translated by lorenzo chiesa

ANGELS

1

Perhaps, in modernity, nothing has had so much been written about it, and with so little perspicacity, as angels. Their image, at the same time beautiful and exhausted, thoughtful and efficient, has so deeply penetrated not only the daily prayers and liturgies of the Occident, its philosophy, literature, painting, and sculpture, but also its day-dreaming, subcultures, and the *Kitsch*, that even a merely coherent comprehension of the topic seems out of question. And when, in the twentieth century, the angel forcefully re-emerges in Rilke's *Elegies* or in Klee's paintings, in Benjamin's *Theses* or in Corbin's gnosis, his gesture does not appear to us today to be any less enigmatic than that of the seraphims who, in the *etoimasia tou thronou* of Palaeochristian and Byzantine basilicas, seem to protect in silence the empty throne of glory.

The situation is completely different if we open the treatises on angels of the Church Fathers and of the Scholastics, from the Pseudo-Dionysius to Alain of Lille, from Bonaventure to Thomas Aquinas, from Dietrich of Freiburg to Eiximens. Here angelology finds its proper place in the economy of the divine government of the world, of which angels are ministers. Not only is Thomas's most extensive treatment of angelology an integral part of the section of the *Summa theologica* dedicated to the government of the world, but the very names of the angelic hierarchies coincide to a large extent with the terminology of power: "Domination, principalities, powers [*potestà*], thrones"; not only is the Pseudo-Dionysius's treatise on angels entitled *Of Celestial Sacred Power* ("sacred power" being the original meaning of the word "hierarchy"),[1] but the very hierarchies of worldly power, both ecclesiastical and profane, regularly present themselves as an imitation of the angelic. The very terminology of modern public administration finds its first formulation in the "offices," "ministries," and "missions" of the celestial functionaries studied by angelology: the concept of hierarchy is an invention of the Pseudo-Dionysius, and the term "ministry" first assumes its modern meaning of "set of functionaries and offices" in a letter by Saint Jerome, in which he asks: "When did God create thrones, dominations, powers, angels and all the celestial ministry [*totumque ministerium coeleste*]?"[2]

2

From this perspective, the distinction between angels and bureaucrats tends to become blurred.

Celestial messengers are organised into offices and ministries just as worldly functionaries assume angelic qualities in their turn and, in the same way as angels, become capable of curing, enlightening, and perfecting. Moreover, because of an ambiguity that characterises the history of the relation between spiritual and secular power, the paradigmatic relation between angelology and bureaucracy runs in both directions. Sometimes, as in Tertullian's and Athenagoras' writings, the administration of the worldly monarchy provides the model for the angelic ministries, whereas in other cases, as in the writings of Clement of Alexandria and Thomas Aquinas, the celestial bureaucracy furnishes the archetype of the worldly.

It is true that the theological tradition distinguishes between two aspects or functions of angels: a properly "governmental" or administrative one, and one of "assistance," in which they contemplate and glorify God (in Dante's words, "contemplative blessedness" and the "blessedness of governing the world"[3]). But one of the essential results of our investigation of the genealogy of government is that these two functions are the two sides of the same governmental machine, which we can call, respectively, "economy" and "glory," "government" and "kingdom." In this sense, each angel is twofold: the ecstatic choirs that chant the eternal glory of God in heaven are nothing other than the ceremonial and liturgical other side of the scrupulous winged functionaries who carry out on earth the "historical" decrees of providence. And it is precisely this co-substantiality of angels and bureaucrats that the greatest theologian of the twentieth century, Franz Kafka, perceived with visionary precision, presenting his functionaries, messengers and servants as disguised angels.

3

The difficulty that the Fathers had to confront in the early centuries of the history of the Church amounted to the reconciliation of Marcion's and the Gnostic god, who is foreign to the world, with the demiurge who creates and rules it, of the *deus otiosus*, who does not look after the fate of creatures, with the *deus actuosus*, who provides for their eternal salvation by means of history. More generally, the ineluctable question that Gnosis bequeathed to the three so-called monotheistic religions was the following: is the Divine, the Highest, separated from the world or does he govern it? The Gnostic theological dilemma does not so much concern the opposition between a good and an evil god as that between a god who is foreign to the world and one who governs it.

It is from this perspective that we should understand the decisive character of the function of angels, not only in Christianity but also in Judaism and Islam. In spite of the different configurations assumed by this function in these three religions, in all cases angelology is inseparable from the answer given to the question we posed above and becomes intelligible only in relation to it. Angelology is, in this sense, the most ancient, articulated, and detailed reflection on that particular form of power or divine action which we could call the "government of the world." Each of the three religions answers the Gnostic dilemma in its own way; the anomalies and analogies in the articulation of their respective angelologies correspond to similarities and divergences between the answers they provide.

4

The best testimony to the eminent presence of angels in the Judaism of the late Classical world is the obstinate and constant attempt by the rabbinic literature of that time to reduce and limit their importance: "If trouble comes upon someone, let him cry not to Michael or Gabriel, but let him cry unto Me and I shall answer" (Jerusalem Talmud Berachot 9: 12). The perseverance with which the Rabbis insist on the ephemeral and inconsistent character of angels, on the dependence of angelic liturgy on that of the Israelites, who sing the praise of God each hour while angels only do it once a day (or even, according to some sources, once a year), constitutes an eloquent demonstration of the privileged role played by angels in liturgy and in the life of the communities, which is proved extensively by liturgical sources and apocalyptic literature. Further evidence of this comes from the admonitions of the Pharisee Paul against the "cult of angels" (*thrēskeia tōn angelōn*, Colossians 2: 18) and from his exaltation of Christ "far above all principality, and power, and might, and dominion" (Ephesians 1: 21).

It is well known that rabbinical Judaism later rediscovered and reassessed the meaning of angelology. This rediscovery takes place by means of the function that angels carry out in the government of the world. Whether they are represented as an infinite army, with its "generals," its "chariots," and "camps" corresponding to the seven celestial spheres – as in the *Sefer Ha-Razim* – or as "doorkeepers," as "ministers" who intercede between God and men, as well as tremendous and tireless singers of Glory – as in the literature of the *Heichalot* – the metaphorical register in both cases is that of a sumptuous and immense court, which surrounds the throne of YHWH.

This governmental vocation of angels becomes increasingly more precise and rationalised, and partly loses its visionary character. In the literature of the Karaites, angels are arranged in four classes, hierarchically ordered according to their function: the servants or ministers, the armies, the powers, and the messengers. The old Talmudic motif according to which angels are created every day and, having sung the hymn of praise, are annihilated in the river of fire from which they originated, is now developed in such a way that these instantaneous angels correspond to the act of government; they are intentionally created by God to accomplish a determined act. Each angel is an act of government, and each act of government is an angel.

The equation between angelology and the government of the world is consolidated by the encounter with Aristotelianism. According to Ibn Daud, God exercises his action on the celestial spheres and on the sublunar world through the mediation of angels, who emanate from him and coincide with the separated intellects of the Arabic-Aristotelian tradition. Maimonides himself accepts the Aristotelian principle according to which God governs the world through the separated intellects that move the spheres; however, he specifies that "Aristotle speaks of separated intellects, while we speak of angels." Divine government coincides with angelic mediation to the extent that Maimonides can write that "you will always find God accomplishing an action at the hand of an angel; and one already knows that the meaning of 'angel' is that of envoy, so that anybody who follows an order is an angel." And the identity between angels and the divine government of the world is, for the physician Maimonides, so absolute that the two can be resolved fully in the very virtues and faculties that form and govern the living body: "God put in semen a formative function that gives a figure and a specification to the limbs, that is, the angel [. . .] Each faculty of the body is an angel and even more so the powers disseminated in the world . . ." From the perspective of the divine government of the world, physiology is an angelology.

If, in the rationalism of philosophers, angels seem to lose all autonomous consistency by being identified with divine action, rabbinical Judaism has never become unaware of a substantial extraneousness between angels and God. Nowhere does this extraneousness appear more strongly than in the book of Enoch, which, according to Scholem, originated in the context of rabbinical Gnosticism. The patriarch Enoch is here transformed into the archangel Metatron, the Prince of the Face, a sort of personification of angelic power. Identified, according to the Talmud, with the "angel of YHWH" (from Exodus 23: 21) and "bearing the same name as his Lord," Metatron is the only angel who sits on a throne in the presence of YHWH. He concentrates in himself so many powers and functions that seeing him is what possibly inspired the apostasy of Elisha Ben Abuyah (Acher in the well-known episode from the Talmud about the four rabbis who enter Paradise). Such apostasy should be interpreted in this context as a fall into Gnostic ditheism. Metatron tells us that

> when Acher came to behold the vision of the chariot and lay his eyes upon me, he was afraid and trembled before me [. . .] When he saw me seated upon a throne like a king, with

> ministering angels standing beside me as servants and all the crowned princes of the angelic realms surrounding me, he opened his mouth and said: "Yes, there are two powers in heaven."[4]

It is surprising that here the angelic-demiurgic power of government is opposed to God as a different power (if not as a different God). The fact that shortly after Metatron is punished "with sixty fiery lashes of fire" (like in the analogous episode of the fustigation of the angel Gabriel in the Talmud – *Yom.*, 77a) is a further sign of the extraneousness and almost virtual rivalry between the angel and YHWH in Judaism. The angel here is the form in which the Gnostic demiurge is subjected to the divine government of the world.

5

The motif of angels called "elements" (*stoicheia*) proves that the archaic figure of the angel as a demiurgic power was constitutively linked to the cosmos. This theme, already present in paganism in the *stoicheiokratores theoi*, the gods who govern the elements, which Simplicius mentions in his commentary on *De coelo*, establishes such a close connection between the elements of the cosmos and the angels that the latter are defined simply as *stoicheia*. An early and decisive occurrence of this can be found in Paul (Galatians 4: 3; 4: 9):

> Even so we, when we were children, were in bondage under the elements of the world [. . .] But now after you have known God, or rather are known by God, how *is it that* you turn again to the weak and beggarly elements, to which you desire again to be in bondage?

An exegetical tradition, already present in Marius Victorinus and in Jerome, identifies these elements of the world with angels: "Some believe that those [elements] are angels, who preside over the four elements of the world, and it is necessary that, before believing in Christ, we are governed by their arbitrariness."[5] Angels-elements also appear in Clement of Alexandria ("the elements and the stars, that is, the powers that govern them") and in Origen, according to whom elements ultimately designate the angels that are presupposed by them. From this perspective, it is possible to put forward the hypothesis that angels are originally cosmic powers which the celestial God needs to subject in order to govern the world.

6

In Christianity, the dualism between a god who is foreign to the world and a demiurge who rules over it is reconciled by transferring it within divinity. The Trinity is the apparatus by means of which God not only takes creation upon himself but also, through Christ and his incarnation, the redemption and government of creatures. This means that Christianity introjects the angelic power into God himself, and turns the government of the world into a divine figure. It is not surprising, then, that the first elaboration of the Trinitarian paradigm between the second and third centuries (more specifically, in the works of Irenaeus, Hippolytus, and Tertullian) takes the shape of an *oikonomia*, that is, an activity of management and government that the Father entrusts to the Son. Clement of Alexandria clearly expresses this essential solidarity between the Trinity and the economy of redemption when he writes that "being done away with providence [*pronoia*], the economy of the Saviour appears to be a myth [*mythos phainetai*]." In other words, the Trinity is not a mythology, a family story like that of the pagans: it is immediately an economy, a cure and government of the world.

The strong solidarity between Christology and angelology follows from this. Not only are angels the instruments of the economy of salvation, but Christ himself is initially presented as an angel, or rather, in Epiphanes' account, "as one of the archangels, but superior to them." In Malebranche's providential theology, Christ, as head of the Church, still appears as the executive head of a *machina mundi* of which God is the supreme legislator; in this function, he is compared with angels and unhesitatingly defined as – even if by now it is only a metaphor – "the angel of the new law." It is significant that this theme of the *Christos-angelos*, to which, in the wake of Werner's monograph, Corbin has drawn the attention of scholars, is the object of a fierce polemic among the Church Fathers. If Christ is an angel and not a God, then the Trinitarian apparatus, which was founded on the introjection of the angel into divine life, not only cannot work but also threatens the unity of the divine. In spite of the resolute elimination of this angelic nature of the Son by means of the doctrine of the *homousia*, the angelological origin of Christology will continue to act in the history of Christianity as an atheological drift that tends to replace the primacy of the eternal being with the historical economy of salvation, the immanent Trinity, defined by the unity of substance, with the economic trinity, which is essentially praxis and government.

From this follows the ambiguity of angelology in Christianity and the necessity to integrally transform it into a bureaucratic-executive structure of divine providence by firmly confining the angelical ranks to the governmental machine. This transformation is fully accomplished by Scholasticism, in which the treatises *De gubernatione mundi* coincide *de facto* or *de jure* with those on angels. Perhaps, the specifically ecclesial and governmental consistency that defines Christianity with regard to Islam and Judaism also follows from this. It is within the reflection on angelic hierarchies as a model of ecclesiastic hierarchies that, starting with the Pseudo-Dionysius (whose work should not be read, following the equivocation that has dominated its reception in the West, in a mystical way, but as an attempt to found the sacredness of power and of ecclesiastic hierarchies on the Trinity and the angelic hierarchies), the first legitimisation of the Church as a "worldly" structure of the government of souls takes shape. The fact that, on the basis of the celebrated passage from Paul (Colossians 1: 18; 2: 10) on the "cephalic" character of Christ (Christ as the head of the angels and "of the body, that is, of the Church"), the power of the Church (and even any power) was founded on Christ and the pope defined as his "vicar," eloquently demonstrates the essentially governmental meaning of Christology.

From the perspective of the government of the world, Christology and angelology, the Messiah and the angels are inseparable and remain so until the Last Judgement, when the history of salvation is accomplished and they are left, literally, with nothing to do. In Christian theology, the paradigm of the government of the world is, as a matter of fact, essentially finite. After the Last Judgement, when all the elected will be taken to Heaven and all the damned sent to Hell, every activity of government ceases and the angelic hierarchies are deprived of all their functions. Of all but one: Glory, which they will continue to tirelessly sing to God for ever. Glory is the form in which the angelic-governmental function survives itself.

7

In Islam, the *tawhīd*, the affirmation of the absolute unicity of God, and the ensuing polemic against the Trinity and the very possibility of Christ having a hypostatic unity composed of a human and divine nature determine the framework in which angelology can carry out its task. The Ash'arite thesis, which in the end prevailed in the Sunni *kalām*, about God's unceasing operation in every event, about God as the single author of every action – good or bad – of man – whose freedom is thus reduced to the "acquisition" of that which in any case he cannot produce – further conditions the very possibility of an angelic government of the world. This does not mean that, in Islam, the function of angels as messengers and assistants of God is not present; however, the focus here is on the function of glory, in which angels appear as the "precursors of each and every act of cult," who "extol" and "worship" God at all times without getting tired.

"Heaven creaks," claims one of the sayings of the Prophet, "and it has the right to creak. There is not in it the space for as much as four fingers without an angel prostrating his forehead there." A sermon of "Alī, the prince of believers," quoted by Rāzī in his angelological treatise, distinguishes between the angels who are prostrated, and never bow, those who bow, and never raise their back, those who draw up, and never break their rank, and those who praise, and never tire themselves of glorifying. The very obedience and fear ("fearing him they tremble") that make it possible for them to scrupulously fulfil their duties as envoys among men are an integral part of their worshipping ("all angels listen to him and obey, they are always in the process of worshipping him").

However, even here, the story from the Koran about the dissatisfaction of the angels with the creation of Adam, and about Iblīs's – the most powerful among them – refusal to worship him, are witness to the irreducible extraneousness of angels with regard to the divine. If, as attested to by some sources, Iblīs corresponded according to this aspect to Enoch-Metatron, the problem concerning the origin and nature of angels emerges in the Islamic tradition especially in the discussions on the superiority of men (and therefore of prophets) over angels (as in the treatises of Rāzī and Tabarī). The opposition between angels – who appear as the representatives of the power of creation – and prophets – who represent the power of the Imperative (or of salvation) – corresponds once again to the Gnostic opposition between a creative and a redemptive God. There are traces of Gnostic dualism even in the most rigorously monotheistic religion.

8

If the government of the world is still today in the hands of the Christian Occident (even if we do not know for how long), this is certainly not unrelated to the fact that Christianity is the only one of the three monotheistic religions that has turned the government of the world into an internal articulation of divinity and that has, thus, divinised angelic power. A "government of men" is not possible in Judaism, in which the angelic function remains somehow foreign to God, even though it is subjected to him; or in Islam, in which God intervenes directly at all times and in all particularities in the course of events. The heterogeneity of political models and the resistance (especially evident in Islamic countries) to the acceptance of external paradigms corresponds to the difference between the respective angelologies, according to their more or less stressed governmental inflection and the different ways of realising it. For a government to be possible it is in fact necessary to have the articulation and the coordination of a bipolar machine made of an immanent angelic power, which operates in particularities as executive, and a transcendent divine power, which acts as a universal legislator. This machine is, in Christian theology, providence, which, by identifying itself with the economy of salvation, binds in an articulation that is equally meticulous and imposing the luminous and immutable plan of the divine mind (*ordinatio* or *providentia generalis*; Malebranche, and Rousseau after him, will speak of "general will") with the untiring, detailed, and seedy intervention of angelic emissaries (*executio* or *providentia specialis*; Rousseau, following Malebranche, will speak of "particular will," or "public economy," or even "government"). In this sense, even if this claim will surprise many, providence, with its angels-bureaucrats, is not the paradigm of absolute power, but of democracy.

9

Angelology is, in this sense, the oldest and most comprehensive reflection on that particular form of power which, in our culture, goes by the name of "government," and which Michel Foucault, starting from the mid-1970s, has tried to define in his courses at the Collège de France. Any attempt to separate angels from their governmental vocation is, in this sense, doomed to fail. In the twentieth century, there have been at least two of these attempts, which are not unrelated: Rilke's poetic project and Corbin's philosophical Gnostic work. In both cases, it is a matter of separating angelology from history, the glorious function of revelation from the obscure and ambiguous function of the government of the world. This is what Rilke means when he writes to Hulewicz that "in the *Elegies* the angel has nothing to do with the angel of the Christian heaven (possibly it has to do with the angelic figures of Islam)." And when, in the same letter, he states that "the angel is the creature in which the metamorphosis of the visible into the invisible that we enact is already accomplished [. . .] that being who is the guarantor of the fact of recognising in the invisible a superior level of reality," the implicit thesis of this affirmation is that angelology – not history – is the place in which the revelation and redemption of the world are fulfilled. For this reason, the *Elegies* are, in the end, hymns in disguise, songs of praise aimed at angels ("Preise dem Engel die Welt . . ."[6]); for this reason, in the *Sonnets to Orpheus*, which contain a kind of exoteric exegesis of the *Elegies*, the task of angels and of men is nothing other than the ceremony of celebration: "Rühmen, das ists!,"[7] "Nur im Raum der Rühmung darf die Klage / gehn."[8] But glory, with its apparatus of liturgies and acclamations, is – as we have shown – precisely the other side of power, the form in which government survives its exercise. And mysticism, both Judaic and Christian, is – at least in one of its aspects – literally only a "contemplation of the throne," that is, of power. Benjamin's opinion about Hoffmansthal, according to which it was Kafka who inherited the legacy of *Lord Chandos's Letter*, and not its author, is, in this sense, also valid for Rilke: the attempt to separate angelology from history in order to transpose the language of poetry into the register of glory closes with a *non liquet*: the lamentation that transforms itself into a celebration is only the ambiguous protocol of reality.

Similar considerations can be made about Corbin's essay "The Necessity of Angelology," which ends, not by coincidence, with a quotation from Rilke. Against Hegel and the theology of incarnation, it is a matter of mobilising the Gnostic theme of the *Christos-angelos* and the Islamic theme of the Shiite *imām* in order to break the connection between angelology and philosophy of history. Redemption is a Gnostic process that does not ever coincide with the level of historical events, for instance, with a revolution, although it can enter into contact with it at some eminent points. But, once again, the theological machine of government is not really neutralised: the hidden and ineffable god, whom the angels have the duty to reveal without offering him any flesh which is not that of an image, is nothing other than the mystical foundation of the power of government, a king who, following a motto dear to Carl Schmitt, "reigns but does not govern."

It is therefore possible that, by joining the figure of the angel with that of history in Klee's *Angelus Novus*, Benjamin prepared for our meditation an emblem which is difficult to dispose of. Kafka must have had something similar in mind when he introduced the functionaries of power as angels (one of them is the "Doorkeeper of the Law" from the parable *Before the Law*) and seemed to recommend, in regards to man's unceasing confrontation with the law, a "long study of the doorkeeper [*jahrelange Studium des Türhüters*]." Angelology and philosophy of history are, in our culture, inextricable, and the possibility of interrupting or breaking their connection – not in the direction of a meta-historical beyond but, on the contrary, towards the very heart of the present – will eventually only open up to those who will have understood it.

notes

1 The title of the work in question is translated into English as *The Celestial Hierarchy*. [Translator's note.]

2 Hieronymus, *Epistolae*, I, 18, 7, in *Patrologiae cursus completus. Series Latina* 22, 365.

3 See Dante, *The Banquet*, Book II, chapter 4, 10–12.

4 Charles Mopsik, *Le Livre hébreu d'Hénoch ou Livre des palais* (Paris: Verdier, 1989) 110.

5 *Patrologiae cursus completus. Series Latina* 26, 371a–b.

6 R.M. Rilke, *Elegies* 9, 53.

7 Idem, *Sonnets to Orpheus* 7, 1.

8 Ibid. 8, 1–2.

alberto toscano

DIVINE MANAGEMENT
critical remarks on giorgio agamben's the kingdom and the glory

"Modernity, having removed God from the world, not only has not exited theology, but it has only, in a certain sense, brought to completion the project of providential *oikonomia*."[1] With these lines, Giorgio Agamben closes his latest, and longest, addition to the *Homo Sacer* project begun in 1995, *Il Regno e la Gloria. Per una genealogia teologica dell'economia e del governo* [*The Kingdom and the Glory: For a Theological Genealogy of Economy and Government*].[2] This magniloquent declaration (there are many in Agamben's work, and in this book especially) encapsulates two key posits of Agamben's research. Firstly, the claim, which drives *Il Regno e la Gloria*, that the Church Fathers, in developing Trinitarian theology, Christology and angelology lay the groundwork for an economic theology of government that remains operative in the current dispensation of Western modernity. Secondly, the idea that the atheism or secularism which nominally characterise contemporary political philosophy – be it liberal, conservative, or Marxist – are surface effects beneath which lie the compulsions of a theological matrix, a "governmental machine" with its roots sunk deep in the Christian past. In other words, the limits and impasses of today's political thought are to be understood in terms of a cunning of secularisation: the apparent disappearance of Christian theology from the commanding heights of politics is but the determinate form taken by the ultimate origination of contemporary political action in the double apparatus composed of a political theology of sovereignty and an economic theology of government and administration – with the latter, as *Il Regno e la Gloria* endeavours to show, playing the key part.

It is not possible in a few pages truly to gauge the significance of Agamben's thesis, or indeed to assay the cogency of his archaeological claims. This essay will merely seek to evaluate the pertinence of the inquiry laid out in *Il Regno e la Gloria* to a radical critique of contemporary politics and economics, in particular its relationship to a Marxian communism which Agamben appears to consider incapable of a truly radical or total critique of the status quo. To do this, I will divide this paper into three sections. Firstly, it will be necessary to get some purchase on what is meant by the "theological genealogy of the economy and government" announced by the book's subtitle. This will involve subjecting to scrutiny Agamben's reliance on a certain understanding of secularisation, of the kind that permits him to declare that modernity merely brings to completion the Christian "economy" of providence, or indeed that Marx's notion of praxis "basically is only the secularisation of the theological conception of the being of creatures as divine operation."[3] As I hope to show, Agamben's work relies on a type of historical substantialism that clashes with his claim to be engaging in a genealogy. Secondly, Agamben's suggestions about the genealogical thread running from Trinitarian *oikonomia* all the way to Smith's invisible hand, and implicitly all the way up to the present, will be contrasted to understandings of the (modern) economy which, being premised on the limitlessness of monetary accumulation, transcends any absorption by a theological genealogy. Finally, we will consider certain aspects of Agamben's archaeological excavations – in particular his delineation of the economic-theological notion of *administration* – and ask whether they might permit a deconstruction of the Marxist reference to communism as the withering away of the state and the shift towards an "administration of things."

on method

Why the turn to a "theological genealogy" of the economy? The origins of Agamben's choice seem to be twofold. On the one hand, there is a desire to prolong the Foucauldian insight into "biopolitics," according to which the primacy of sovereign power is both supplanted by, and recombined with, a government of life – wherein power is not aimed primarily at sheer domination or dealing out death, but at a productive management of individuals and populations. On the other, Agamben takes his cue from a debate between Carl Schmitt and the theologian Erik Peterson, taking his distance from Schmitt's allegiance to the idea of political theology and showing that Peterson himself steps back from admitting the significance to the early Christian theologians of the notion of *oikonomia*. Agamben's study is primarily a painstaking and erudite investigation into the different figures taken by the "economy" in early Christian theology, which all revolve around a basic semantic constellation (a *Sinn*, rather than a *Bedeutung*, as Agamben specifies), in which economy stands for an immanent and "an-archic" management, a generalised pragmatics. In other words, where Foucault had located, beginning in the mid-eighteenth century, the emergence of "governmental reason" in the early discourse of political economy and the concurrent practice of administering the health and productivity of populations,[4] Agamben turns the clock back two millennia, to the writings of Aristotle and Xenophon on *oikonomia*, then to the fate of this notion within the theology of the Church Fathers, beginning with Paul. Defined by Aristotle as the "administration of the house," in contradistinction to the form of collective, public power exercised in the *polis*, in Xenophon

> *oikonomia* is presented as a functional organisation, an activity of management which is not bound to rules other than that of the orderly functioning of the house (or of the undertaking in question). It is this "managerial" paradigm that defines the semantic sphere of the term *oikonomia* (as of the verb *oikonomein* and of the noun *oikonomos*) and determines its progressive analogical broadening outside of its original limits.[5]

As Agamben details, the *oikonomia* is modelled by Xenophon on the organisation that pertains in an army and on a boat. But if the semantic core of the idea of economy is already implanted in Ancient Greek philosophy, why engage in a *theological* genealogy? By this syntagm Agamben evidently intends more than tracking the applications and mutations of *oikonomia* in the ambit of Christian theology.

What is at stake becomes clearer when Agamben comes to the discussion of the place of *oikonomia* in what he calls the providential paradigm and the "ontology of acts of government" that underlies it. As he writes:

> Providence (government) is that through which theology and philosophy try to deal with the splitting of classical ontology into two separate realities: being and praxis, transcendent good and immanent good, theology and *oikonomia*. It presents itself as a machine

> aimed at rearticulating together the two fragments in the *gubernatio dei*, in the divine government of the world.[6]

A theorem from *Seinsgeschichte* – the sundering of being and praxis – is adduced to account for the determining significance of Christian theology in shaping the political and metaphysical horizon of "the West" (a term that Agamben seems to employ without much reflection) up to "our" very own modernity. In this regard, it is the specifically *Christian* fate of *oikonomia*, as the anarchic immanence of a divine government tenuously articulated, via providence, with a transcendent God who "reigns but does not govern," which justifies the theological character of this genealogical investigation. For Agamben, the "providential *dispositif* (which is itself nothing but a reformulation and development of theological *oikonomia*) harbours something like the epistemological paradigm of modern government." In the guise of separation between "legislative or sovereign power and executive or governmental power," the modern state *inherits* "the theological machine of the government of the world." Agamben indicates one of the more troubling dimensions of this inheritance in his beguiling archaeology of the notion of "collateral effects," and the related "collateral damage." As he writes:

> The paradigm of the act of government, in its pure form . . . is the collateral effect. To the extent that it is not directed toward a particular end, but it derives, as a concomitant effect, from a law and general economy, the act of government represents a zone of undecidability between the general and the particular, between the calculated and the non-wanted. This is its "economy."[7]

But by what right does Agamben pass from the insistence of certain conceptual constellations and semantic kernels across different epochs and discursive formations to the overarching conviction that such an archaeological inquiry is of urgent political significance? Note that unlike a historian of ideas or concepts who might wish to track the secret endurance and operative impact of certain thought patterns across periods, Agamben is unconcerned with forms of transmission beyond textual ones. When, for instance, he states that Malebranchian occasionalism transits into Rousseau's conceptions of political economy and popular sovereignty, or that a theological notion of order subtends Smith's invisible hand, the terms of this pernicious theological inheritance are not contrasted to alternative genealogies. Nor does Agamben consider the possibility that the persistence of certain thought forms might be less relevant than their redeployment to radically different ends within incommensurable discursive formations. Nor, finally, is there a serious consideration given to the possibility – abetted, for instance, by the derivation, which Agamben himself shows, of the theological *dispositif* of bureaucracy from the empirical history of empires – that it is not so much the continuity of the theological but the persistence of certain social relations and their imaginaries, which explains the insistence of certain ideas of government throughout such a *longue durée*.

It is symptomatic in this respect that, at the very outset of the book, Agamben haughtily dismisses the theoretical significance of the secularisation debate that pitted the likes of Blumenberg, Schmitt and Löwith against one another in the 1960s – treating it merely as a cloaked struggle over the philosophy of history and Christian theology. For Agamben, secularisation is a strategic gambit, not a historiographic thesis. As a strategy, secularisation – for instance, as famously deployed by Schmitt – involves the polemical reference of political terms to their theological origin. It is here that Agamben introduces a rather mystifying "methodological" term, that of the *segnatura* (signature). Secularisation functions as an element within a science of signatures, that is, a study of "something that, in a sign or a concept, marks it and exceeds it to refer it back to a determinate interpretation or a determinate domain, without thereby departing from the semiotic to constitute a new signification or a new concept."[8] In a recent methodological essay that traces the genealogy of the signature, Agamben cites a study of *Les Mots et les choses* by the Italian scholar Enzo Melandri, where the latter refers to the signature as a "kind of sign within the sign;

it is that index which, in the context of a given semiology, univocally refers back to a given interpretation."[9] Leaving aside for the moment the rather perverse torsion which permits Agamben to turn a term which, as he recognises, Foucault locates in Paracelsus and a pre-Enlightenmental episteme of resemblance, into a notion of which Foucault's own theory of the statement in *The Archaeology of Knowledge* is but an instance, it is worth insisting on what happens to the idea of secularisation once it is treated as "a strategic operator, which marks [*segnava*] political concepts to refer them back to their theological origin."[10] To begin with, such a conviction, accompanied by the rather mystical postulation that only some may "possess the capacity to perceive signatures and to follow the dislocations and displacements that they operate in the tradition of ideas,"[11] means that there is no need to actually gauge the mechanisms that allow for the transition from one discursive field to another, since the very presence of the signature *immanently* refers us back to an origin in the theological field, which accordingly, in a move meticulously dissected in Hans Blumenberg's work, *delegitimates* the political concepts themselves. Political economy, for example, is reduced to a "social rationalisation of providential *oikonomia*."[12] The "theory" of signatures thus seems to engage in what we could call a *reductivist idealism*, a mirror-image of sorts of the much-maligned Marxian reduction of ideal structures to social relations – a materialist move which certain passages of Agamben's book would make rather more plausible than the search for theo-economic signatures, for instance when – referring to the pseudo-Aristotelian treatise *De Mundus* – Agamben shows how the perception of the governmental apparatus of the Persian king would influence the later image of divine hierarchies, as "the administrative apparatus through which the sovereigns of the earth conserve their kingdom becomes the paradigm of the divine government of the world."[13]

But something more problematic is at stake than Agamben's reference to a mode of research, the search for signatures, which so heavily depends on putative personal insight and analogical thinking. This has to do with the idea of a theological *origin*. Behind this reference lies not only Agamben's sympathy towards the Schmittian notion of secularisation but the conviction, mediated by a pervasive Heideggerianism, of a historical-ontological *continuity* which allows one to argue that our political horizon is still determined – and worse, *unconsciously* determined – by semantic and ideational structures forged within a Christian theological discourse. Though Agamben does not straightforwardly embody the apologetic Christian purposes that Hans Blumenberg identifies in the discourse on secularisation – the idea that the conceptual patrimony of the Church was expropriated and misused – he does manifest one of the key aspects of that discourse, the idea of a substantial continuity, without which, we could add, the theory of the signatures becomes inoperative. As Blumenberg writes: "Only where the category of substance dominates the understanding of history are there repetitions, superimpositions and dissociations – and also, for that matter, disguises and unmaskings." Despite his inevitable Heideggerian protestations to the contrary, it is only the idea of an underlying continuity – the continuity of historical-ontological *destiny* – that can allow Agamben, to cite Blumenberg, "to identify the substance in its metamorphoses." Against the idea of a history that is veiled to itself, of secularisation as a kind of spell, which only the man of signatures could dispel, it might be worth reflecting on the suggestion that "there exists a high degree of indifference between a concept and its history."[14]

Along these lines, it is difficult to ignore that the Schmittian and Heideggerian lenses through which Agamben approaches Foucault's methodology lead to a basic and glaring infidelity towards the maxims that orient Foucault's work – above all the Nietzschean and Bachelardian principle of genealogical and archaeological *discontinuity*. As Foucault elucidates in his seminal essay "Nietzsche, Genealogy, History," the notion that there is no semantic conservation, and that genealogy is concerned with dispersed events of heterogenesis and truncated lineages, means that the search for continuities which defines the history of ideas is to be subjected to unsparing critique. Nietzsche's genealogy of morality is accordingly juxtaposed to his friend Paul Rée's *history* of morality. The latter "assumed that words had kept their meaning, that desires still pointed in a single direction and that ideas retained their logic; and then ignored the fact that the world of speech and desires has known invasions, struggles, plundering disguises, ploys." Hence Foucault's reference to the "singularity of events outside of any monotonous finality," as what the genealogist must concern himself with, in a spirit of documentary "restraint" with respect to a history "without constants." Only the "grey" work of genealogy can break with the pious, metaphysical idea that things have a timeless essence, or indeed an unchanging semantic or ontological core, revealing instead "the secret that [things] have no essence or that their essence was fabricated in a piecemeal fashion from alien forms."[15] Not only the origin, but the idea of a disguised "unthought," is repudiated by Foucault for the sake of the discontinuous positivity of an analysis of discourse. As he declares in "L'Ordre du discours":

> The existence of systems of rarefaction does not imply that, over and beyond them lie great vistas of limitless discourse, continuous and silent, repressed and driven back by them, making it our task to abolish them and at last to restore it to speech. Whether talking in terms of speaking or thinking, we must not imagine some unsaid thing, or an unthought, floating about the world, interlacing with all its forms and events. Discourse must be treated as a discontinuous activity, its different manifestations sometimes coming together, but just as easily unaware of, or excluding each other.[16]

But it is precisely the belief in both continuity and concealment which so dominates Agamben's *theological* genealogy, and his interpretation of the redemptive function of archaeology itself. Contrary to Foucault, for whom the duty of genealogy is *not* "to demonstrate that the past actively exists in the present, that it continues secretly to animate the present, having imposed a predetermined form to all its vicissitudes,"[17] Agamben is adamant that the archaeologist re-ascends history against the grain in order to access a historical and anthropological redemption which – in an intriguing reference to Islamic theology – for Agamben *precedes* creation itself. The archaeologist's gesture, far from a grey and patient labour that might allow us, precariously, to think otherwise, is "the paradigm of every true human action."[18] As such a paradigm, archaeology also turns out to be – in a veritable apotheosis of historical substantialism, arguably a by-product of Agamben's peculiar fusion of Heidegger, Schmitt and Benjamin – the only political gesture in a fully unified horizon whose origin and "anarchic" functioning is to be sought in the idea of *oikonomia*, of government as the malleable and endemic *management* and production of "collateral damage." It is also the substantialist thesis which allows Agamben to repudiate as deluded and unreflexive either the fidelity to the watchwords of radical political theory (e.g., Rousseau's general will, another "theological inheritance") or the attempt to revive a secularist critique of religion. As he declares in the appendix of *Il Regno e la Gloria* on the invisible hand, the *oikonomia* of the moderns leaves untouched the concept of government that accompanied the theological model of the government of the world:

> This is why it is senseless to oppose secularism [*laicismo*] and the general will to theology and its providential paradigm, but only an archaeological operation of the kind that we have attempted here which, going back before the split that produced them as rival but inseparable brethren, can dismantle and render inoperative the whole economic-theological apparatus.[19]

the mismeasure of money

In an interview regarding his research on *oikonomia*, which preceded the publication of *Il Regno e la Gloria*, Agamben helpfully summarises the parameters of his inquiry. He describes the Ancient Greek paradigm of *oikonomia* as a "managerial" one, as

> an activity which is not bound to a system of norms nor does it constitute an *episteme*,

> a science strictly speaking, but it implies decisions and dispositions which change from case to case to deal with specific problems. In this sense, a correct translation of the term *oikonomia* would be, as the Liddell-Scott lexicon suggests, *management*.

This semantic core, or *Sinn*, is then transposed, with Clement and Origen, into the initial conceptualisations of history in Christian theology, where history appears as the "mystery of the economy," or, we could say, the mystery of divine management – a management which, as Agamben notes elsewhere with reference to the arguments of Reiner Schürmann, is *anarchic*. Moreover, prolonging his fidelity to the notion of secularisation, Agamben notes that history is accordingly "a mysterious economy, a divine mystery which is the object of Christian revelation and which man must therefore learn to decipher. Hegel (and Marx after him) only pick up this paradigm to finally reveal the mystery." The historical substantialism that we criticised above is evident in such passages, and it also defines Agamben's claim for the political relevance of his archaeological operation. Despite the fact that he will barely touch on modern political economy, and then only through a very brief and tendentious treatment of Rousseau and Smith, it is implied throughout that some kind of thread unites the "anarchic" providentialism of Christian theology with our own capitalist predicament. Indeed, we could hazard that this millennial detour which references Foucault's own (far more modest) turn to the analysis of economic thought (neoliberal governmentality in particular), with its attenuation of the centrality of a Schmittian *political* theology, is one way for Agamben to respond to the obvious critique of the *Homo Sacer* series as one which entirely ignored *capitalism* as a singular form of (bio-)power and an inexorable constraint on the various modalities of sovereignty and law. Though Agamben does mildly protest that it would be "excessive" to say that he is trying to "reconstruct the essence of capitalism," he does nevertheless declare that: "We can only understand the triumph of the economy today if we think it alongside the triumph of the managerial paradigm of theological *oikonomia*." It is in the (empty) concept of order – that link, or "signature" that connects the immanence and transcendence, praxis and being, which the emergence of theology itself sundered – that Agamben observes an "essential presupposition [which] binds ancient economy to modern economy" via the theological paradigm.[20]

It would be interesting to consider what an attention to their theological precursors would have to tell us about modern concepts of economic order – for instance Hayek's notorious neoliberal ontology of "spontaneous order." It would, of course, be necessary not to rely on the philosopher's capacity to divine theological signatures. Astronomy, as in the case of Smith's "invisible hand," might, for instance, be a more pertinent source-domain for notions of order. In any instance, the possibility of asymmetry, disconnection or indifference between causes and effects should inspire in us a certain genealogical restraint. After all, that the contemporary concept of order may be related back for a number of its features or for its very structure to mediaeval theological elaborations says very little about its functioning or its validity – unless, of course, we have already accepted the idea that we remain caught within a theo-economic apparatus which has been in place ever since the putative collapse of the unity of being and praxis, a unity whose recovery would somehow spell redemption.

But what is more damning to Agamben's claims for the political urgency and epochal depth of his archaeological operation is the absence of the other paradigm of "economic" behaviour which Aristotle in particular defined, only in order to ward it off as a potential threat to the order and stability of the polis: *chrematistics*, the science of monetary accumulation, circulation and interest that is opposed to the managerial stability of the paradigm of *oikonomia*. However "anarchic" the managed order heralded by *oikonomia* might be, it is itself threatened by another kind of anarchy, that of money as a "real abstraction" that threatens to obliterate any stable measure, any standard of judgement, any principle of order. Marx noted this encounter of philosophy with the scandal of accumulation in an important note to the first volume of *Capital*. It is worth quoting in full.

> Aristotle opposes Œconomic to Chrematistic. He starts from the former. So far as it is the art of gaining a livelihood, it is limited to procuring those articles that are necessary to existence, and useful either to a household or the state. "True wealth (ο αληθινοζ πλοντοζ) consists of such values in use; for the quantity of possessions of this kind, capable of making life pleasant, is not unlimited. There is, however, a second mode of acquiring things, to which we may by preference and with correctness give the name of Chrematistic, and in this case there appear to be no limits to riches and possessions. Trade (η καπηλικη) is literally retail trade, and Aristotle takes this kind because in it values in use predominate) does not in its nature belong to Chrematistic, for here the exchange has reference only to what is necessary to themselves (the buyer or seller)." Therefore, as he goes on to show, the original form of trade was barter, but with the extension of the latter, there arose the necessity for money. On the discovery of money, barter of necessity developed into *καπηλικη*, into trading in commodities, and this again, in opposition to its original tendency, grew into Chrematistic, into the art of making money. Now Chrematistic is distinguishable from Œconomic in this way, that "in the case of Chrematistic circulation is the source of riches (ποιητικη χρηματων...δια χρηματων διαβοληζ). And it appears to revolve about money, for money is the beginning and end of this kind of exchange (το νομισμα στοιχειον και περοζ τηζ αλλαγηζ εστιν). Therefore also riches, such as Chrematistic strives for, are unlimited. Just as every art that is not a means to an end, but an end in itself, has no limit to its aims, because it seeks constantly to approach nearer and nearer to that end, while those arts that pursue means to an end, are not boundless, since the goal itself imposes a limit upon them, so with Chrematistic, there are no bounds to its aims, these aims being absolute wealth. Œconomic not Chrematistic has a limit... the object of the former is something different from money, of the latter the augmentation of money... By confounding these two forms, which overlap each other, some people have been led to look upon the preservation and increase of money ad infinitum as the end and aim of Œconomic." (Aristoteles, "De Rep." ed. Bekker, lib. l. c. 8, 9, *passim*)[21]

Chrematistics, by transgressing the natural order of needs and positing a *limitless* accumulation thereby presages the principle of capitalism as *self-valorising*, but also as the annihilatory and dissolving force depicted in *The Communist Manifesto*. One way of formulating this distinction in terms already encountered via Agamben is that chrematistics, in having money as both origin and end, threatens to generate an entirely *unmanageable* economy, and thus to mine the order of needs subtending the polis, as well as the very capacity for judgement itself. As Eric Alliez notes in an important analysis of this point of encounter between Aristotle and Marx, chrematistics introduces a "time of dislocation," a "world crisis" into Aristotelian politics and cosmology, "replacing the social unity of need, the natural referent of the monetary sign, with interest." Chrematistics is a

> hybrid science... which distinguishes itself from *oikonomia* governed by use-value in that circulation becomes the source of an unlimited monetary wealth, "money is the beginning and end of this kind of exchange": M-C-M.′ A science of money whose bad infinity haunts the organicity of the political body by deregulating the postulate of exchange between equivalents.[22]

More recently Chris Arthur has tried to show, on the basis of the thesis that Marx's systemic dialectics of capital bears a considerable isomorphy to Hegel's logic, that money instantiates both a true infinity, since "it returns to itself in its circuit," and a spurious, or bad infinity, since "capital is embarked on the escalator of accumulation and cannot get off." The restlessness of money as capital within the "spiral" of accumulation, means, echoing Aristotle's fears, that neither limit nor measure are capable of giving it a stable shape.[23] The very form of value is such that

> the good and bad infinities get all mixed up; because here we have a Being-for-Itself furthering itself through its own otherness;

> but whose peculiar essence is to be pure abstraction of quality (use value), namely quantity (value); hence the movement is limitless, it must always go on, for its return to itself always fails to close with itself because its very essence is boundlessness. Marx says: "Capital as such creates a specific surplus-value because it cannot create an infinite one all at once; but it is the constant movement to create more of the same". So a particular capital never measures up to its concept and is compelled to throw itself into ever more twists of the spiral of accumulation.[24]

This very cursory treatment of the question of chrematistics, of an economy of limitlessness and accumulation, suggests that Agamben's theological genealogy is incapable of providing much insight into the (value) forms that determine the (dis)order of the contemporary economy. While *Il Regno e la Gloria* does provide a rich archive for a study of the Christian prehistory of "management" as an increasingly endemic principle of social organisation, it is entirely mute – arguably because of Agamben's banal Heideggerian prejudices about the place of labour and productivity in the Marxian critique of political economy – about the "anarchic" order of capital accumulation, and thus has nothing to say about the constitutively unmanageable economies (*chrematistic*) that management (*oikonomia*) seeks to govern. The discontinuity and asymmetry between the economic and the chrematistic, or between management and accumulation, also indicates the poverty of trying to perpetuate the tired idea of Marx's thought as a "secularisation" of some cloaked and damning theological content. The signatures just aren't there. Neither capitalism nor Marx's theory thereof can be encompassed by the notion of *oikonomia* and its genealogies, theological or otherwise, and it does not suffice to combine political theology with economic theology to overcome the shortcomings of Agamben's work as a tool for politically thinking the present.

the administration of things

As a coda of sorts, I want to end by considering the relevance of a theme in *Il Regno e la Gloria*, the putative theological origins of bureaucracy and administration, to a possible critique of the Marxian communism which Agamben, as a good left-Heideggerian, regards as hampered, alongside the whole of the political theory of the West, by its "theological inheritance."[25] It would not be difficult to imagine a prolongation of Agamben's argument on the economies of administration into a critique that would overlap with the many that have been levied against the communist thesis of the "withering away of the state" as a post-political utopia (or dystopia) of transparent planning. For Agamben, the very modern notion of administration – which one could discern in texts such as Engels's "On Authority" or much of Lenin's post-1917 production – is bound up with the providential apparatus, with a governmental machine that links the transcendence of a plan to the immanence of a government that is always a government over collateral effects. "The modern state," he writes,

> effectively inherits both the aspects of the theological machine of the government of the world, and presents itself both as welfare-state [*stato-provvidenza*] and destiny-state. Through the distinction between legislative or sovereign power and executive or governmental power, the modern state takes upon itself the double structure of the governmental machine.[26]

What's more, the modern state, as Agamben notes, is also, if we follow the theological signatures, the very model of *hell* – it was indeed the indefinite continuation of *oikonomia* with no possibility of salvation that in Christian theology marked the fate of the damned. So, is the reference to the "administration of things" a sign that Marxism, too, is caught up in the bureaucratic *ministerium* first formulated in Christian angelology, that it, too, carries with it a hellish hierarchical order?

As Hal Draper has noted, the idea of the passage from the government of men to the administration of things, which originated with Saint Simon and has often been cited by both anarchists and Marxists, can certainly carry a rather hellish omen:

> This is usually taken to be a laudable sentiment meaning the abolition of the rule

> of man over man; but in fact Saint Simon's highly despotic schemes showed him to mean, in his governments, something quite different: the administration of men as if they were things.[27]

And when Engels writes that under communism "public functions will lose their political character and be transformed into the simple administrative functions of watching over the true interests of society,"[28] one might hear echoes of Agamben's *gubernatio dei*,[29] or Foucault's own characterisation of *Polizeiwissenschaft* in terms of "a system of regulation of the general conduct of individuals whereby everything would be controlled to the point of self-sustenance, without the need for intervention."[30] But before we rush to consider communist politics as yet another secularisation, we need to consider the determinate manner in which it is pitted, not against sovereign power or bureaucratic management, but against the kind of economic coercion wielded by capitalist accumulation, that is by the value form. In this regard, it is worth considering how the problem of the "administration of things," rather than merely a conduit for a bureaucratic substance going back to Christian angelology, is there to answer a profound political-economic problem: what would a (communist) society be beyond the real abstractions of Capital and the state? In other words, what would it be to organise society without money as its measure and the privatisation-expropriation of public power by class interest? It is here that a key "economic" question, that of equality, comes to the fore. I will conclude with a brief discussion of the problematisation of equality by Marxism, in order to show the political difference which is introduced by thinking our predicament not in terms of *oikonomia* but in terms of capitalism, not in terms of theological genealogy but of historical materialism.

It is only in terms of its attempt to move beyond the forms of social order and measure borne by the paradigms of *oikonomia* and chrematistics that we can make sense of the critique of (political and economic) equality within communist thought. Consider the *Critique of the Gotha Programme*, and the commentary on that document in Lenin's *State and Revolution*. Faced with a truly "economistic" theory of justice (the social-democratic ideal, pushed by the likes of Lassalle, that equality signifies "fair distribution," "the equal right of all to an equal product of labour"), Marx retorts that the notion of equality implied by this distributionist vision of communism is still steeped in the very abstractions that dominate bourgeois society, that is, it is still bound to the unstable relation between abstract political equality and limitless accumulation under the aegis of the value form which marks out the specificity of capitalism. Speculating about a communist society that *emerges from* capitalist society – and is thus not just its negation but its *determinate* negation – Marx notes that the abrogation of exploitation and the capitalist appropriation of surplus value would *not yet* end the forms of injustice that inhere in the domination over social relations by the abstraction of value. In a nascent communist society, distribution is still "governed by the *same* principle as the exchange of commodity equivalents: a given amount of labour in one form is exchanged for the same amount in another." In other words, a certain economy still functions as a constraint that no amount of genealogy or archaeology could redeem us from.

Equality in such an embryonic, transitional communism is still beholden to the domination of a standard, *labour*, which is itself the bearer of inequalities – of capacity, productivity, intensity, and so on. The equal right so blithely invoked by the social-democrat is thus "in its content one of inequality, just like any other right," since "a right can by its nature only consist in the application of an equal standard" to *unequal* individuals. In other words, a political and philosophical notion of equality as a right, founded on the idea of an abstract and universal measure or standard, still bears the birthmarks of a form of social measurement based on the value of labour, on its "economy." In Lenin's gloss,

> the mere conversion of the means of production into the common property of the whole of society . . . *does not remove* the defects of distribution and inequality of "bourgeois

right" which *continues to dominate* in so far as products are divided "according to work."

In light of these statements, we can reflect that communism, and its horizon of "administration," is the determinate and not the simple negation of capitalism. The communist problem of equality is the problem of an equality, to quote Lenin, *without any standard of right* – which is to say an equality that does not perpetuate the inequalities generated by the domination of social relations by the measures of value, by the labour-standard in particular, which pertain to capitalism. Such a "non-standard" equality can only be thought as an outcome of a revolution and transition that would not only abrogate capitalism's chrematistic creative destruction but also the abstract forms of right and sovereignty that determine equality within bourgeois society.

But in undoing the pact between the measuring measurelessness of money and the liberal standards of abstract rights articulated with concrete oppressions, does this perspective transcend the horizon of Agamben's theo-economic machine of government? Of course, the concern with production and labour, so abhorrent to Agamben, means that classical communist thinking, for all of its concern with the liberation of time and even play, is alien to the "sabbatical" political anthropology advocated by Agamben. Human essence qua "ensemble of social relations" is not simply "inoperative," impotential. The preoccupation with real needs and material constraints, as well as with the resistances of nature, means that some form of "economic" thinking, of governing and ordering and distributing resources, is inescapable outside of the purely religious horizon of redemption. In that regard, regardless of its theological genealogy, a dimension of bureaucracy, though not of hierarchy, accompanies any communal pursuit, though the struggle to produce polyvalence is aimed at forestalling the reification of human relations into orders of function and specialisation. But there is a specifically Marxian sense of the inoperative which attention to the question of equality brings to the fore, and which arguably has a concrete, if utopian, force that Agamben's liminal and messianic anthropology lacks. The inoperative is here a procedure and not an essence, a political practice and not what lies on the other side of a catastrophic distinction between living under and beyond a given apparatus or *dispositif*. Rather than *either* affirming the principled equality of human beings *or* promising their eventual levelling, communist "equality" involves creating social relations in which inequalities would be rendered inoperative, no longer subsumed as unequal under an equal standard or measure of right. In other words, the challenge of communism is that of producing a politics without an *arche* that would not merely be a form of governance dominated by an absent principle – an absence that, following Agamben's suggestions, would be filled in by the spectacle of glory. Its immanence would no longer be mined by an absent God and its inscrutable ministers. But this other administration only makes sense if we circumvent the mirage of an anthropology of redemption for the sake of a thinking not of popular sovereignty but of collective or trans-individual power, something that Agamben's Heideggerian ban on a "metaphysics of the subject" and on "humanism" prevents us from doing.

notes

1 Giorgio Agamben, *Il Regno e la Gloria. Per una genealogia teologica dell'economia e del governo* (Vicenza: Neri Pozza, 2007) 314. This paper will limit itself, in an admittedly truncated way, to the discussion of *oikonomia*, leaving to one side the important, and in some ways more persuasive, analysis of the spectacle of political glory proposed by Agamben, as well as its links to the theme of a political anthropology of the inoperative – that dimension which both economy and glory, in their twin functioning, allegedly obscure, or even repress.

2 Agamben has promised a fourth and final volume, on "forms of life." The English translation of this text was published after the writing of this article. See *The Kingdom and the Glory: For a Theological Genealogy of Economy and Government*, trans. Lorenzo Chiesa with Matteo Mandarini (Stanford: Stanford UP, 2011).

3 Ibid. 106.

4 See especially the first lecture, dated 10 January 1979, in Michel Foucault, *Naissance de la biopolitique. Cours au Collège de France, 1978–1979* (Paris: Seuil/Gallimard, 2004) 3–28. Agamben disregards the importance of the questions of liberalism and of the self-limitation of government to Foucault's account.

5 Ibid. 32–33.

6 Ibid. 157. The "an-archic" Greek sources of such a notion of providential government are to be found, among others, in the writings of the Aristotelian commentator Alexander of Aphrodisias, for whom, according to Agamben,

> what is essential is not so much the idea of a pre-determined order, as much as the possibility of managing disorder; not the inexorable necessity of fate, but the constancy and calculability of a disorder; not the uninterrupted chain of causal connection, but the conditions of maintaining and orienting effects which are in themselves purely contingent. (140)

7 Ibid. 158.

8 Ibid. 16.

9 Giorgio Agamben, "Teoria delle segnature" in *Signatura rerum. Sul metodo* (Turin: Bollati Boringhieri, 2008) 61.

10 Ibid. 68.

11 *Il Regno e la Gloria* 16.

12 Ibid. 310.

13 Ibid. 96.

14 Hans Blumenberg, *The Legitimacy of the Modern Age*, trans. Robert M. Wallace (Cambridge, MA: MIT P, 1983) 9, 15, 21.

15 Michel Foucault, "Nietzsche, Genealogy, History" in *Language, Counter-Memory, Practice: Selected Essays and Interviews*, ed. Daniel F. Bouchard (Ithaca, NY: Cornell UP, 1977) 139, 142. Agamben briefly discusses this piece in his essay on philosophical archaeology in *Signatura rerum*, but he seems to ignore the profound challenge that the very idea of genealogy poses to his own research, forcing it instead into a redemptive vision of an "ontological anchoring" (111) quite alien to Foucault's text. Foucault states his allegiances to a kind of relational nominalism even more strongly in a 1982 interview with Paul Rabinow, where he declares:

> Nothing is fundamental. That is what is interesting in the analysis of society. That is why nothing irritates me as much as these inquiries – which are by definition metaphysical – on the foundations of power in a society or the self-institution of a society, and so on. These are not fundamental phenomena. There are only reciprocal relations, and the perpetual gaps between intentions in relation to one another. (See Michel Foucault, "Space, Knowledge, and Power" (1982 interview) in *Power: Essential Works of Foucault 1954–1984*, by Michel Foucault; ed. James D. Faubion (London: Penguin, 2001) 356)

16 Michel Foucault, *The Archaeology of Knowledge and The Discourse on Language* (New York: Pantheon, 1972) 229.

17 "Nietzsche, Genealogy, History" 146.

18 *Signatura rerum* 108.

19 *Il Regno e la Gloria* 313.

20 On the medieval, post-Aristotelian development of order as a political and metaphysical paradigm, see *Il Regno e la Gloria*, especially 99–105.

21 Karl Marx, *Capital*, vol. I [1887], translated from the third German edition by Samuel Moore and Edward Aveling (London: ElecBook, 1998) 219–20.

22 Eric Alliez, *Les Temps capitaux. Tome I: Récits de la conquête du temps* (Paris: Cerf, 1991) 30–32.

23 The only "measure" for capital being, as Arthur notes, a *rate* of accumulation – that is, a form of measure deeply destructive of the kinds of measures envisaged by an Aristotelian *oikonomia*.

24 Christopher J. Arthur, *The New Dialectic and Marx's Capital* (Leiden: Brill, 2004) 148–49.

25 On the theme of left-Heideggerianism, see Matteo Mandarini, "Beyond Nihilism: Notes towards a Critique of Left-Heideggerianism in Italian Philosophy of the 1970s" in *The Italian Difference*, eds. Lorenzo Chiesa and Alberto Toscano (Melbourne: re.press 2009) 29–48. Agamben's study of glory could, of course, be linked to the debate over political religions and cults of personality as they affect the history

of communism. The thesis that the exacerbation of glorification is a sign of the failure of a transformation in the practice of government is quite perspicuous, but it's not evident that it needs the ontological and anthropological commitments (the idea of the human as a "sabbatical" or inoperative creature) lent to it by Agamben.

26 *Il Regno e la Gloria* 159.

27 Hal Draper, "The Death of the State in Marx and Engels," *Socialist Register* (1970) 282.

28 Engels, "On Authority" qtd in Draper 290.

29 Consider especially the following passage: "Governing means allowing for the production of the particular concomitant effects of a general 'economy' that would remain in itself entirely ineffective, but without which no government would be possible" (160). It would be intriguing to think of the economic idea of the Plan along these lines . . .

30 "Space, Knowledge, and Power" 351.

jelica šumič

GIORGIO AGAMBEN'S GODLESS SAINTS

saving what was not

In his rich and insightful commentary on Agamben's work, "Giorgio Agamben: The Discreet Taste of the Dialectic," Antonio Negri praises Agamben for his attempt to formulate a radically immanentist conception of redemption. He points out, however, that this thoroughly novel vision of redemption, according to which "this world, this being in which we are immersed, is also a source of possibility," is rooted in an ontological choice, "a kind of Heideggerianism," Negri observes, since for Agamben, just as for Heidegger, "being is given and destinal."[1] Owing to this "unresolved condition" of his thought, Negri argues that there are "two Agambens": one that "recognizes and denounces the fact that the state of exception... has come to involve all structures of Power," the other "traversing it with feverish utopian anxiety, grasping its internal antagonism" (DTD 118–19). Yet precisely to the extent that Agamben is unable or unwilling to separate the given from the destinal aspect of being, Negri continues, his ultimately "nihilistic conception of being" hardly leaves any room for "a radically inventive revolutionary activity" – except in the guise of "passive marginal resistance" (DTD 118–21). At this point, it becomes possible to see why Agamben's position, which can recognise resistance "as passivity rather than as rebellion, represented by Bartleby, rather than Malcolm X, by *homo sacer* rather than by the slave or the proletariat" (DTD 123), is for Negri singularly unproductive. This is because behind "the heroism of the negative," Negri argues, we only discover "the return to nothingness, to the destinal insignificance of being, to the marginality of refusal" (DTD 124). Ultimately, Agamben's universe, in which all productive power is attributed to sovereign power alone, reminds Negri of that of an obscure philosopher of the seventeenth century, Arnold Geulincx. The latter envisioned "a world that the transcendent divine cause had so totally invested that it closed down all spaces for freedom and production" (DTD 125).

At the end of *Il Regno e la Gloria*, in what appears to be an indirect response to this critique, Agamben celebrates Bossuet's theodicy that seeks to "reconcile human freedom with the divine government of the world" by claiming that "God... wishes for all eternity that man be free, and not only potentially but in the actual and concrete exercise of his freedom."[2] Indeed, what makes Bossuet's work interesting for Agamben is precisely this extreme providentialism since it allows us to see how "theology can resolve itself into atheism, and providentialism into democracy, because *God has made the world just as if it were without God and governs it as though it governed itself*" (RG 314). Interestingly enough, Agamben's booklet *The Coming Community* ends on a practically identical note, with a tiny displacement, however: instead of envisaging God having created the world as if there were no God, as Bossuet did, Agamben boldly states that precisely this "godless" world, at once transient and irreparable, that is, "the world – insofar as it is absolutely, irreparably profane – is God."[3]

If the focal point of my essay, its vanishing point, as one could say somewhat facetiously, is this pivoting of perspectives as presented in Bossuet's extreme providentialism and Agamben's own identification of the "irreparably profane" world with God, this is because this dialectic of immanence and transcendence throws into relief what is at stake in Agamben's singular conception of redemption.

Paradoxically, it is Agamben's commitment to redemption that explains what is at stake in his rejection of "immanentism." The reason for this is the mutation of the present regime of mastery, which, having as its structural principle the generalisation of exception, succeeds in creating through this very lawlessness an interminable status quo, immune to all change. This is particularly clear in Agamben's claim that what characterises our contemporary situation is an auto-suspension of its structuring principle, "a suspension of the law as such."[4] Outlining some of the fundamental features of the state of exception in which we allegedly live today, Agamben notably insists on "an absolute indeterminacy between inside and outside."[5] With the progressive abolition of the distinction between observance and transgression of the law, the state of exception marks the emergence of a regime where "there is no 'outside' of the law" (TR 105), a regime of the not-all in the proper sense of the word in so far as, in such a regime, the exception to the law, instead of being excluded, is generalised. This is why the state of exception presents itself as a state "where everything becomes possible" (TR 106), as a state, moreover, that tends towards "transforming itself into a lasting model of governance" (SE 17). For what is paradoxical about the regime founded on the generalised exception and suspension of the law, a world in which the law is made to coincide entirely with the lawlessness, is that the regime, instead of breaking down, keeps running.

The eternalisation of the existing state of affairs provides us with a plausible key to identifying the difficulties of contemporary immanentist theories of resistance, such as Negri's, in finding a way out of the present impasse. Indeed, once it is revealed that the principle of the functioning of the master discourse is "dysfunction," since in the present state of exception "the law is in force only in the form of its suspension" (TR 105), thereby ensuring the eternalisation of the thus created state of exception, one should tease out the implications of Benjamin's analysis of the baroque state of exception in order to identify the threats inherent in the self-generating immanentism of the state of exception. Drawing on Benjamin's analysis, Agamben argues that the baroque "knows an *eskhaton*, an end of time," yet precisely because it "remains immanent to this world" it "knows neither redemption nor a hereafter" (SE 56). Ironically, the baroque "godlessness," as Benjamin insisted, instead of opening a path to "a redeemed hereafter," could only end in a catastrophe. But precisely for this reason, nothing is more urgent in the contemporary state of exception, according to Agamben, than to "halt the machine" (SE 87) and, in so doing, to find a way to wrench from it a potential for transformation. No doubt, Agamben's initial idea of a way out of the current impasse is to remind us not to take the existing state of exception as necessary and therefore immune to change. However, it is also clear that any attempt at changing the current state of affairs requires different means and strategies of resistance, rather than simply seeking to subvert the law.

In his attempt to address the question of redemption in our time, to a certain extent Agamben follows Adorno: for both thinkers, redemption is concerned with the restoration of the possible which is given only in retrospect as an always-already missed opportunity.

According to Adorno, it is the "despair of the world" that, paradoxically, "guarantees to us that the hopelessly missed things exist."[6] Hence the task set for thought is none other than to restore the possibility that dwells within these "hopelessly missed things." This can only be achieved by turning our attention not to things as we find them solidified in concepts but rather to things as they are in their "becoming." Thus, to see things in their becoming or, to use Adorno's proper formulation, to read things "as a text of their becoming" (ND 52), is to glimpse what he calls "the possibility of which their reality has cheated the objects and which is nonetheless visible in each one" (ND 52).

This becoming of things that is, strictly speaking, given only in retrospect, through cracks and fissures in their appearance, is no doubt a fantasy, a utopia, says Adorno. Nonetheless, this utopia yields hope. This yielding of hope is all the more paradoxical since it is grounded in a fantasy, staging not what a thing could have become but rather what it has failed to become. It is this "messianic" aspect of redemption, latently present in Adorno, that has been taken up and further developed by Agamben in his theory of redemption conceived in terms of the saving of the past, more exactly, of saving "what was not."

But in order to understand Agamben's theory of redemption and the special place that God or "godlessness" has in it, it is first necessary to highlight the importance of the idea of the irreparable. Much of Agamben's discussion of redemption turns indeed on the "irreparably profane" status of the world. If "the Irreparable is that things are just as they are [. . .] consigned without remedy to that way of being," as Agamben insists, then it is in a sense logical that "the possibility of salvation" begins with the recognition of the irreparability of "how the world is" (CC 89). Agamben himself emphasises the paradoxical nature of redemption as he envisions it. Redemption, he argues, is not to be conceived of as "an event in which what was profane becomes sacred and what was lost is found again." Rather, redemption consists in "the irreparable loss of the lost, the definitive profanity of the profane." This is because "we can have hope only in what is without remedy" (CC 101). In order to make this point clear Agamben continues:

> That things are thus and thus – this is still in the world. But that this is irreparable, that this *thus* is without remedy, that we can contemplate it as such – this is the only passage outside the world. (The innermost character of salvation is that we are saved only at the point when we no longer want to be. At this point, there is salvation – but not for us.) (CC 101)

Or to put it somewhat differently, it is precisely because "the taking place of things does not take place in the world" that there is a hope for salvation. And it is in this sense that Agamben can claim that "Utopia is the very topia of things" (CC 102).

In light of this conception of redemption, the only exigency that thought is capable of assuming fully, according to Agamben, is to rescue "the capacity to remain faithful to that which having perpetually been forgotten, must remain unforgettable. It demands to remain with us and be possible for us in some manner" (TR 40). Exigency, thus conceived, refers to something "which remains unforgettable [. . .] the shapeless chaos of the forgotten" and which, precisely for having been forgotten, is still "at work within us" (TR 39–40). It is in light of a threat of the return of this forgotten non-dead that Agamben conceives of exigency as a sort of protection, a prophylactic, against the return of the forgotten, or, more properly, against the return in the real of that which has been foreclosed from the symbolic. Indeed:

> If, however, we refuse to respond, and if, on both the collective and individual level, we forgo each and every relation to the mass of the forgotten that accompanies us like a silent golem, then it will reappear within us in a destructive and perverse way, in the form Freud called the return of the repressed, that is, as the return of the impossible as such. (TR 40–41)

It is in view of such a defence against a possible destructive return of the forgotten past that the relation between the possible and the real must be rethought. Thus, against Leibniz, who stated that "every possibility demands to exist, to become real," the gist of exigency, for Agamben, lies in a retroactive production of the possibility of that which is presented to us in its immutable givenness. Hence the inversion of Leibniz's formulation: "Each existent demands its proper possibility, it demands that it become possible. Exigency consists in a relation between what is or has been, and its possibility" (TR 39). The exigency of restoring the unforgettable qua unforgettable thus necessarily implies a radical transformation of the past.

For what is at stake here is nothing less than the exigency to restore the contingency of the past or, in Agamben's words, the "redemption of what has been" (TR 41). This redemption, he argues, consists in nothing less than the restoration, after the fact, retroactively, as it were, of the possibility, not of that which will take place but, paradoxically, of that which has taken place. It is only under this reading that the possible "does not precede reality; rather, it follows it" (TR 39).

However, for such a retroactive production of the possible to be possible, the subject's intervention is required. This is a curious intervention, to be sure, since the subject, confronted with the forced choice either to identify itself with the role assigned to it by the Other or to rebel, opts for the impossible, the "choice of the worst," as exemplified by Bartleby's infamous formula: "I would prefer not to." We are presented here with an "impossible" choice, yet one that renders both options, to obey or to disobey, equivalent, thereby shattering the very framework that made this either/or possible. But this "third," properly inexistent choice, a choice that must first be created, invented, requires an exorbitant price.

The subject will pay the price for creating this impossible choice by retroactively putting into question its subjectivisation. For what is crucial in this destitution of the subject is a wager that is placed, in Lacan's terms, not on the father but on the worst (*parier du père au pire*), which is to say, on that which is from the outset lost: the object *a*. Or, as Agamben himself rightly remarks, the subject that takes up the messianic task to be faithful to the unforgettable, "settles its differences once and for all with presumed identities and ensuing properties" (TR 41) of the subject. Indeed, if "[t]he coming of the Messiah means that all things, even the subjects who contemplate it, are caught up in the *as not*, called and revoked at one and the same time," as Agamben insists, then the crucial point to note here is that there is "no subject to watch it or decide. The messianic vocation dislocates and, above all, nullifies the entire subject" (TR 41; translation modified). At this point, Agamben surprisingly joins psychoanalysis, since salvation, as he sees it, is centred on a new subjective figure, one that takes upon itself "the assimilation to what has been lost and forgotten" (TR 41).

This assimilation to a sort of waste product is close to the position of the analyst, at least in Lacanian psychoanalysis. For the analyst to be efficacious in the analysis, "the subject of the unconscious," Lacan maintains, must "take him as the cause of the subject's own desire."[7] But the price to be paid for occupying the position of the cause of the subject's desire is the analyst's subjective destitution. Thus, it could be said that the analyst's transformation into a cause of the desire of another subject, the analysand, is "paid for" by the analyst's conversion into an object. Moreover, to be operational in the analysis, which is only possible through the incarnation of the excessive leftover, that which does not count and which, for that reason, finds no place in the given order, the analyst must be willing to exit from human society, in a word, to be a dropout of humanity. Ultimately, to be true to his role, the analyst must "truly want to be shit."[8]

It is because the analyst can become the object *a*, trash, that Lacan compares him to the saint. In doing so, Lacan emphasises the ambiguity of the saint's status. As Benveniste pointed out, two terms are required to express the singularity of sainthood (*sacer/sanctus*), to signify the positive and negative aspects of this mode of existence, at once "charged with divine presence" and "forbidden to human contact."[9] Lacan's surprising identification of the analyst with the saint follows from the negative side of the word *sanctus*, in the double sense of "consecrated to the gods and charged with an indelible stain, august and cursed, worthy of veneration and cause for horror."[10]

Crucial for our discussion here is that, in a situation in which it seems that there is no option left, Lacan puts forward a solution which consists, ultimately, in identifying the position of the subject, not with the agent or the producer but with the product or, more precisely, with what remains after production, what is left over, with trash. This also explains why the position of the saint is presented by Lacan as a privileged site of resistance to the hegemonic discourse, which, for him, is that of the capitalist. As a model of self-positioning in spaces in which the distinction between the inside and the outside is abolished by the dominant discourse itself, sainthood as practised by the analyst would, according to Lacan, succeed in jamming the machine of capitalism that feeds on the want-to-enjoy. If sainthood is capable of interrupting the insatiable "more" of the capitalist drive for growth, as Lacan maintains, this is because the saint is one who refuses to produce, but instead persists in a certain modality of passivity or inoperativity, and therefore uselessness.

There are some undeniable points of convergence between the analyst-saint's "doing nothing" and the *désoeuvrement* of man – a Kojèvian notion taken up by Agamben and used to describe a certain modus of passivity that would designate the "non-acting action" proper to the role to be played by the analyst in an analysis. Seen in this way, the saint on which Lacan models the analyst's refusal to be useful, to surrender to the demands of capitalism, might be understood as a strategic posture of resistance: as an "extimate positionality" vis-à-vis the position of the subject required and modelled by the dominant discourse, the saint is situated at the level of what is left after the completion of dis-identification, i.e., at the level of that which cannot be represented, the infamous object *a*.

In this regard, it is perhaps no accident that much of Agamben's discussion of the present possibility of a break with the existing state of exception is centred on the figures of extimacy, such as the messianic subject or *homo sacer*. Similarly characterised, just like Lacan's analyst-saint, in the negative terms of a dropout of humanity, *homo sacer*, as Agamben remarks, "belongs to God in the form of unsacrificeability and is included in the community in the form of being able to be killed" (*Homo Sacer* 82).[11] At the same time, Agamben locates in Paul's epistles the object *a* as the only tenable position, which he describes in the following terms: "We are made of filth of the world, the offscouring of all things" (1 Cor. 4: 13). A messianic subject is not one who, from his privileged position, contemplates the world such as it would have been if it were saved. On the contrary, "the one who upholds faith in what is lost" (TR 41) is one who "contemplates salvation only to the extent that he loses himself in what cannot be saved" (TR 42). At stake in the messianic destitution of the subject, as Agamben himself rightly points out, is the force of suspension, a putting into parentheses of every point that could serve as a support for any identity or worldly vocation.

It is precisely the destructive power of this suspension that Agamben reveals and explores in his "Bartleby, or On Contingency."[12] In this respect, another image could corroborate our reading, one used by Agamben himself, although in a different text and for a different purpose, namely the *tabula rasa*. According to Agamben, "the image of the writing tablet on which nothing is written functions precisely to represent the mode in which potentiality exists" (P 224). It would perhaps be more appropriate to say, as Agamben himself observes, that Bartleby is less to be viewed as a "*tabula rasa*" than as a "*rasura tabulae*" (P 243), the gesture of "*far tabula rasa*," of "making a clean sweep," that very gesture which, by wiping the tablet clean, creates an "emptiness," or, to be more precise, "not even a space, but the site of a place."[13] Interestingly, Agamben does not shy away from exalting the liberating power of the indiscernibility between potentiality and impotentiality. It is this indiscernibility which appears, in Agamben's eyes, as the subject's most precious treasure, its agalma, which sustains it in its lack of being. Indeed, if "the human being exists in the fracture between the living being and the speaking being, the human and the inhuman," as Agamben remarks, it follows that "the human being is the being that is lacking to itself and that consists solely in this lack and the errancy it opens."[14] What restores the human being as "a potential being" (RA 134), or, to be more precise, what restores the poetic status of man in the world, the potentiality that we are, is not the world, which confronts us in its unfathomable givenness, but a perspective which opens us to the transformation of the world that is presented to us precisely as given and immutable into a space in which the possibility of the world being otherwise than it is, indeed, the possibility of another world, can be brought to presence.

It is at this juncture that we can return to Agamben's own "immanentist" conception of redemption. Rejecting three possible readings of salvation (in messianic terms), that is to say: the reading that envisions salvation as an unrealisable or "impotential" utopia, where one is condemned to pretend that salvation is real simply by acting as if salvation were true; the reading where the emphasis is on a blind belief in the ineluctable necessity of redemption; and, finally, the reading which, by deeming the world unsaveable, resentfully turns against it, thus opening up the path for Nietzsche's solution: to "will backwards," that is to say, to transform every "thus it was" into a "thus I willed it" (P 267), Agamben insists that salvation should be considered "as an exigency for this world now." The messianic subject is not one who contemplates "the world as though it were saved" (TR 42), but one who "knows that in messianic time the saved world coincides with the world that is irretrievably lost, and that [...] he must now really live in a world without God" (TR 42). Thus, Agamben's conception of redemption may well be radically immanentist, but, in order to save this idea of an immanent redemption, he has to turn to a specific reading of the messianic, in particular the messianic time as a creation of another time (and space) within the existing historical time, a time in which the past, that is, what has been, is saved integrally, but, and this is a paradoxical aspect of the messianic redemption, in its retrievability.

If Agamben considers the potentialisation of the past, its liberation, as a way out from the here and now, this is not only because it allows for the fulfilled to become unfulfilled – through the restoration of its potentiality not to be – and for the unfulfilled to become fulfilled – through its potentiality to be. Crucial, for him, is rather that mankind, by liberating the past, which paradoxically consists in rendering it irretrievable, unsaveable, liberates itself from the past. It could also be said that Agamben's project of potentialisation aims at bringing into being another world, which can be attained only through an inversion of perspective which enables us to perceive the world not in its givenness but in its potentiality not to be. What such an operation implies is nothing other than a return to a kind of zero degree, a point at which the potentiality to be equates with the potentiality not to be.

Messianic "recapitulation," the fulfilment of time, could then be seen as a periodic recapitulation of what has taken place, a break with the necessity–impossibility nexus that subtracts the possibility of change from the powers of mankind. Here a sort of inversion of the perspective is invited: in times in which everything is possible, in an era in which sovereign power itself is structured as a generalised exception, when every oppositional activity can be co-opted by established structures of power, in order to interrupt the eternalisation of the existing state of emergency, the solution, according to Agamben, consists in a move from the end of time, this being the time in which we are already living in our future, because no radical change is to be expected from the future, to the time of the end, a time of the potentialisation of the world, that is literally extracted, stolen time, as it were, from the worldly time.

What is crucial in this shift of perspective, from which the world appears to us in its potentiality not to be, i.e., its radical contingency, as a world that could be otherwise than it is, is that, at the subjective level, it requires a corresponding mutation which makes it possible for the subject to re-discover "that there is no essence, no historical or spiritual vocation, no biological destiny that humans must enact or realize," yet that this dizzy liberation from teleology does not mean that humans

> are simply consigned to nothingness and therefore can freely decide whether to be or not to be, to adopt or not adopt this or that destiny [. . .] There is in effect something that humans are and have to be, but this something is not an essence nor properly a thing: *It is the simple fact of one's own existence as possibility or potentiality.* (CC 40)

Put otherwise, instead of seeking to accomplish some definite task or goal, the subject must be nullified. Indeed, it is only through the destitution of the subject that man's capacity to be pure potentiality can be restored.

Attributing all transformative force to sovereign power alone, Agamben's solution, whose ultimate aim is to restore contingency at the heart of necessity, consists in directly valorising the "not happening" or rather the "nothing of happening" in order to consign change to a radical transformation in the subjective status, achieved by means of an operation of disidentification that aims, to use Agamben's vocabulary, at revoking all vocations. Agamben can recognise resistance only in terms of potentiality, which is to say, as passivity or inoperativeness, since, for him, "the potential welcomes non-Being, and this welcoming of non-Being *is* potentiality, the fundamental passivity" (P 182). To the extent that the potentiality that characterises human beings is primarily the potentiality of not doing something, the subject, here, is conceived as a place where the ceaseless operation of declassification, disidentification, is effected – Bartleby being the model or paradigm of such a subjective stance in so far as the latter allows the subject to become nothing other than the pure potentiality to be or not to be.

The characterisations of the subjective stance in terms of inoperativeness can be seen as an attempt to move beyond the deadlocks of the end of time in so far as such a stance involves "a suspension of time" achieved through the only possible action at the disposal of contemporary subjectivity, an action *à la* Bartleby, an anticipatory figure: to opt for non-being, or more exactly, for the potentiality not to be. His act ("I would prefer not to" . . .), in effect, consists in a mere taking place of the place. He turns himself into a place, an empty place, this being the only place which sovereign power cannot recapture. However, for this place to be preserved, maintained as a place, nothing should take place therein. His act, instead of constituting an event, in its subversive force, prevents all events from happening. Indeed, Agamben's Bartleby can be seen as a guardian of the non-event. Ultimately, rather than risking the danger of falling prey to a bad infinity, Agamben seeks to think a final event.

Thus, in contrast to Badiou, who thinks events as time-breaking and/or inaugurating ruptures, Agamben's main preoccupation is with the event of the end. In light of this, one can also understand why the politico-ethical solution advocated by Agamben essentially consists in saving the past: not something particularly worthy of being remembered, but the past in its whatever character, as it were. The world can only be saved if its being-thus in the smallest details is preserved. What is saved, then, is not some break-inaugurating moment, a moment of "eternity," as Badiou would have it, but the banality of the being-thus. It is precisely for this reason that the world can only be "saved" as irreparable, which is to say, ultimately, as absolutely unsaveable. The salvation and therefore the change of the world consist, in the final analysis, only in assuming its radical contingency.

A true change consists simply in a parallax view, a shift of perception: to see the world as including its potentiality not to be. Yet this change, Agamben insists, minimal as it may appear, is nevertheless extremely difficult to accomplish. In some radical sense, humankind is incapable of achieving it; hence, in order to attain this perspective, the Messiah must come or, at least, Bartleby. If Deleuze, as Agamben observes, is right in calling Bartleby "a new Christ," this is not because his aim is to "abolish the old Law and to inaugurate a new mandate." Rather, "if Bartleby is a new Messiah, he comes not, like Jesus, to redeem what was, but to save what was not" (P 270). If there is something Christ-like in Bartleby, if he can, despite everything, be compared to a saviour, this is because he descends to the deepest level of Leibniz's "Palace of Destinies," in order to reveal "the world in which nothing is compossible with anything else, where 'nothing exists rather than something'" (P 270). In Badiou's vocabulary, we could say that Bartleby reveals the inconsistency of being-multiple, an impossible point of the real before Being is localised in any being-there whatsoever, before any world whatsoever can take shape. Whereas this impossible-real, according to Badiou, can only irrupt to the surface of a world through a rare, unpredictable event, Agamben, on his part, presents it as a result of the subjective destitution. Following Badiou's reading of Mallarmé, we are tempted to state that Bartleby, in this respect at least, is perhaps closer to Mallarmé's destitute master who, hesitating to throw the dice, "comes to equate gesture with non-gesture."[15] The master, on this reading, is not only the one who decides or chooses but also the one who, in the very act of decision or choice, suspends it, thus rendering choice and non-choice equivalent.

Placing key emphasis on the subjective rather than the historical aspect of a break with the current impasse, Agamben's solution to the problem of the way out is, thus, articulated through a recasting of the relationship between the past and the present. As has been remarked by a number of commentators, Bartleby has no past or future; indeed, his only time is *the present, the time of the now*, to borrow Benjamin's term, a curious temporality since it cannot be situated "outside of chronological time either" (TR 64). Instead, it can be represented as "a caesura which, in its dividing the division between two times [chronological time and *eschaton*], introduces a remainder into it that exceeds the division" (TR 64). In this regard, Agamben argues that, in every representation a speaking being makes of time, it seems as if another time is produced with respect to the chronological time. However, this "other time" is

> not a supplementary time added on from outside to chronological time. Rather, it is something like a time within time – not ulterior but interior – which only measures my disconnection with regard to it, my being out of synch and in non-coincidence with regard to my representation of time, but precisely because of this, allows for the possibility of my achieving and taking hold of it. (TR 67)

It could then be said that Bartleby's time, the time in which his act, his operation of potentialisation is accomplished, is precisely the "operational time" (TR 68) Agamben attributes to the messianic subject, working within the chronological time and transforming it from within. How are we to understand this necessity for a "time within time," if for the messianic subject, as Agamben claims, salvation has already taken place? How are we to conceive of this "paradoxical tension between an already and not yet that defines the Pauline conception of salvation" (TR 69)? Agamben's answer is the following: although salvation has already been accomplished, it still requires, "in order to be truly fulfilled," "an additional time" (TR 69). More precisely:

> Whereas our representation of chronological time, as the time *in which* we are, separates us from ourselves and transforms us into impotent spectators of ourselves – spectators that look at the time that flies without any time left, constantly missing ourselves – messianic time, an operational time in which we take hold of and achieve our representations of time, is the time we ourselves are, and for this very reason, it is the only real time, the only time we have. (TR 68)

This curious time, "the time we ourselves are," marks, according to Agamben, "that innermost disjointedness within time through which one may grasp time and accomplish it" (TR 72). Obviously, messianic time only starts ticking once the (messianic) event has taken place, a singular event to be sure because it is itself generative of time: for such an event – which, because of its unfathomable nature, we fail to grasp – to be fully accomplished, time is needed which cannot be measured and whose duration is in principle endless, the time that we propose to call, borrowing Lacan's term, the "time for comprehending." It should be noted, however, that, whereas for Lacan, who was the first to break up time into three qualitatively different dimensions (the instant of the glance, the time for comprehending, and the moment of concluding),[16] the time for understanding is necessary, without being a sufficient condition for the subject to accomplish his or her act, i.e., to reach a decision in an aporetic situation, Agamben, by contrast, arrests the subject in the time for understanding.

This is because the possibility of "reading" the world differently, i.e., from the perspective of its potentiality, the possibility opened up by the messianic event whereby the world's irreparability, its ultimate "being thus" is viewed as a mark of its radical contingency, is not a preparation for a change that will take place in the future: it is the only change that man as a poetic agency can expect and wait for. Hence, instead of envisaging messianic time as a "transitional time," a "delay" which tends to prolong itself indefinitely, we should rather consider it as an operational time which consists in the subject's turning towards its here and now in order to accomplish the ceaseless operation of disidentification. It is precisely because it cannot be situated within a linear temporality of past, present, and future that this time of the now is, as such, the location in which action, the hollowing out of the assigned identities, functions, or symbolic mandates take place.

This also explains why the messianic subject, Bartleby, is arrested, blocked, as it were, in the "time of the now," i.e., at the point of the suspension of time, in order to be able to effect his act, that of the de-activation of identifications assigned to him by the socio-symbolic Other. The result of the messianic act is not a new creation – it is rather a decreation. From such a perspective, Bartleby can be seen as someone who turns himself into an utterly irreducible remnant, the sole guardian and guarantor of the empty place destined for "the experience of taking place in whatever singularity" (CC 24). But the price to be paid for this operation of exposure of every singularity to its being-thus, its being whatever, is that the subject himself remains blocked, suspended on the sole act he can effect: *I would prefer not to*, an act which, in so far as it must be repeated again and again, imprisons the subject in a kind of tense-less space created by this very act. There is then no new present for the subject, only this "time of the now" as the time of the end. Indeed, for Agamben, precisely because the messianic vocation "calls for nothing and to no place . . . it may coincide with the factical condition in which each person finds himself called" (TR 24). But this is possible on condition that what is at stake here is nothing less than "*the revocation of every vocation*" (TR 23) in so far as the messianic vocation, working from within the factical condition, "hollows it out, nullifying it in the very gesture of maintaining and dwelling in it" (TR 24). Hence, the messianic act is capable of revoking the factical condition, of undermining it "without altering its form" (TR 24). Agamben maintains that "the expropriation of each and every juridical-factical property . . . under the form of the *as not*" results in "the new creature," which is "none other than the use and messianic vocation of the old" (TR 26). Agamben's paradigmatic case of such a use of the generic potentiality is the experiment accomplished by the Franciscans which consisted in creating "a space that escaped the grasp of power and its laws, without entering into conflict with them yet rendering them inoperative" (TR 27). The gist of this operation resides precisely in the fact that the factical vocation "is not replaced by something else, but is rendered inoperative" (TR 28). Or put somewhat differently, although juridically unchanged, the factical condition, in so far as it is subtracted from the law, "remains a place of pure praxis, of simple 'use'."

It is in light of this "use," which renders a given factical identity or symbolic mandate inoperative, that one can re-examine the operation effected by Bartleby. For Agamben, it is not an accident that Bartleby is a "law-copyist," "a scribe in the evangelical sense of the term." His "renunciation of copying" must be read precisely in relation to the Law, i.e., as "a liberation from the 'oldness of the letter'." In this sense, Deleuze is right in calling Bartleby "a new Christ," since Bartleby "comes to abolish the old Law" and to inaugurate a new mandate. However, Bartleby is not simply a new addition to Christ. If he is "a new Messiah" this is only because "he comes, not like Jesus, to redeem what was, but to save what was not" (P 270). This is why Bartleby's mission is not "to bring a new table of Law but... to fulfil the Torah by destroying it from top to bottom" (P 270). Or, to be even more precise, what Bartleby aims at, ultimately, is nothing other than the restoration of the impossible-real of the law, its originary lawlessness, which is to say meaninglessness, by reducing it to a heap of "unordered letters" (P 167), the reverse side of its symbolic pendant: "The writing tablet... on which nothing is written" (P 166).

At this point, it becomes possible to see why man's creation is actually decreation. If "Scripture is the law of the first creation [...] in which God created the world on the basis of its potential to be," and if, as a consequence, every letter of the law "signifies both [...] what was and what could not be," Bartleby's act, "the interruption of writing," marks what Agamben calls "the second creation," one which, in summoning all potential not to be, consists, actually, in nothing less than in creating "a point of indifference between potentiality and impotentiality" (P 270). Viewed from this perspective, this second creation, which is "neither a re-creation nor an eternal repetition," is revealed to be "a decreation in which what happened and what did not happen are returned to their originary unity in the mind of God, while what would have not been but was becomes indistinguishable from what could have been but was not" (P 270).

Hence, it could be said that it is only through a true act of decreation, a subtractive act, to be sure, that the mark of contingency in every creature is revealed. If decreation, as Agamben tells us, "takes place where Bartleby stands" (P 271), we must ask: what exactly is this place where "the actual world is led back to its right not to be; [where] all possible worlds are led back to their right to existence"?

Here, Lacan's famous formula, "The word is the murder of the Thing," can help us to illuminate this singular position of the subject: if the signifier "creates" by breaking the biunivocal correlation of the word and the thing, if the word does not represent the thing but can only attain a meaning by being articulated to another word, this means that the signifier already de-realises or un-realises the world. The act of the signifier is precisely an act of decreation, rendering indistinguishable that which exists from that which does not exist. If the signifier itself empties all reference, what could then be Bartleby's decreation? Consider Bartleby's formula: "I would prefer not to." As Deleuze correctly observed in his reading of "Bartleby,"[17] Bartleby may well use signifiers, yet he does it in a very peculiar way since his formula is destined primarily to cut the link between words and things, between S_1 and S_2, leaving S_1 all alone, in sufferance, in eternal anticipation of the other signifier that would give it a meaning. But this formula is itself possible because Bartleby occupies the place of an internal exclusion in relation to language. Put otherwise, only for a subject that is outside discourse, discourse, which for Lacan is precisely the social bond, is nothing but a fraud, a make-believe. Bartleby's decreation, in short, can only be effected from the autistic position of the subject who refuses to be caught in any social bond whatsoever, who wants nothing, yet prefers not to, who treats signifiers as fragmented bodies, without any reference whatsoever to a symbolic order.

It is here that we can see what is subversive, really revolutionary, in the act of decreation. Accomplished by the subject for whom there is no distinction between the real and the symbolic, indeed, by a subject for whom the symbolic is, as such, the real, the act of decreation brings into question the Other, the guarantor of the link between words and their references. If Deleuze is right in claiming that Bartleby, "even in his catatonic or anorexic state" is not the "sick man" but rather the "Medicine-Man, the new Christ or the brother to us all,"[18] this is because only from the position of the inexistence of the Other – this being, according to Deleuze, the position of the schizophrenic – the symbolic can appear, for other speaking beings, those who believe in the Other and live by its laws, and who use the symbolic as a defence against the real, as mere semblance. From such a perspective, Bartleby's act can be viewed less as an act that decreates the created (i.e., the symbolised universe) than as one that decreates the decreation, a decreation to the second power, as it were, because such an act of decreation aims at revealing the generalised semblantification at work in the symbolic order itself. If the schizophrenic position, a position outside discourse, suits well the revolutionary who strives to unbind the existing social bond in order to postulate a different basis for a community, beyond identifications, beyond functions and places, this is because it embodies the liberating potential, as well as its risks. For Lacan, as is well known, "not only can man's being not be understood without madness, but it would not be man's being if it did not bear madness within itself as the limit of his freedom."[19] Indeed, it is only from such a position of extimacy in relation to the social link that "the law of our becoming" can be formulated: "The unsoundable decision of being in which human beings understand or fail to recognize their liberation, in the snare of fate that deceives them about a freedom they have not in the least conquered."[20]

notes

1 Antonio Negri, "Giorgio Agamben: The Discreet Taste of the Dialectic" in *Sovereignty and Life*, eds. Matthew Calarco and Steven DeCaroli (Stanford: Stanford UP, 2007) 117. (Hereafter cited parenthetically by page number as DTD.)

2 Giorgio Agamben, *Il Regno e la Gloria: Per una genealogica teologica dell'economia e del governo. Homo Sacer, II, 2* (Milan: Neri Pozza, 2007) 314. (Hereafter cited parenthetically by page number as RG.)

3 Idem, *The Coming Community*, trans. Michael Hardt (Minneapolis: U of Minnesota P, 1993) 89. (Hereafter cited parenthetically by page number as CC.)

4 Idem, *The State of Exception*, trans. Kevin Attel (Chicago: U of Chicago P, 2005) 56. (Hereafter cited parenthetically by page number as SE.)

5 Idem, *The Time that Remains*, trans. Patricia Dailey (Stanford: Stanford UP, 2005) 105. (Hereafter cited parenthetically by page number as TR.)

6 T.W. Adorno, *Negative Dialectics*, trans. E.B. Ashton (New York: Continuum, 1973) 372. (Hereafter cited parenthetically by page number as ND.)

7 Jacques Lacan, *Television*, trans. Jeffrey Mehlman (New York: Norton, 1990) 15.

8 Jacques Lacan, "Discours à l'école freudienne de Paris" in *Autres écrits* (Paris: Seuil, 2001) 275.

9 Emile Benveniste, *Vocabulaire des institutions indo-européennes. 2: Pouvoir, droit, religion* (Paris: Minuit, 1969) 179.

10 Ibid.

11 Giorgio Agamben, *Homo Sacer: Sovereign Power and Bare Life*, trans. Daniel Heller-Roazen (Stanford: Stanford UP, 1998) 82.

12 Idem, "Bartleby, or On Contingency" in *Potentialities: Collected Essays in Philosophy*, ed. and trans. Daniel Heller-Roazen (Stanford: Stanford UP, 1999). (Hereafter cited parenthetically by page number as P.)

13 Idem, *Idea of Prose*, trans. Michael Sullivan and Sam Whitsitt (Albany: SUNY P, 1995) 33.

14 Idem, *Remnants of Auschwitz: The Witness and the Archive*, trans. Daniel Heller-Roazen (New York: Zone, 2002) 134. (Hereafter cited parenthetically by page number as RA.)

15 Alain Badiou, *Handbook of Inaesthetics*, trans. Alberto Toscano (Stanford: Stanford UP, 2005) 50.

16 Jacques Lacan, "Logical Time and the Assertion of Anticipated Certainty" in *Écrits*, trans. Bruce Fink (New York: Norton, 2006) 161–75.

17 Gilles Deleuze, "Bartleby, or: The Formula" in *Essays Critical and Clinical*, trans. D.W. Smith and M.A. Greco (London and New York: Verso, 1998) 68–90.

18 Ibid. 90.

19 Jacques Lacan, "Presentation on Psychical Causality" in *Écrits*, trans. Bruce Fink (New York: Norton, 2006) 144.

20 Ibid. 145.

boštjan nedoh

KAFKA'S LAND SURVEYOR K.

agamben's anti-muselmann

introduction

Numerous critics have recently pointed out a general problem that traverses the entirety of Giorgio Agamben's work. In his analysis of the politico-ontological fates of various figures, such as, for example, Bartleby the Scrivener, *homo sacer*, and the *Muselmann*, Agamben does not leave room for any consistent and operative theory of the subject, even though he himself frequently relates these figures precisely to the possibility of a "new politics" that would suspend and transcend the structures of sovereign power and biopolitics.[1] The critics in question identify two main features of his thought as the cause for the absence of a theory of the subject. On the one hand, Agamben aims at developing the field of the Aristotelian ontology of potentiality, which is most convincingly condensed in Bartleby's famous phrase "I would prefer not to."[2] Agamben argues that the syntagm with which the scrivener answers his employer when he persistently assigns him various kinds of work tasks needs to be understood precisely as the "potentiality not to."[3] The key consequence of this negative determination of potentiality lies precisely in the fact that, as potentiality, it is not realised in actuality. There is an absence of action. We are confronted here with the basic problem of Agamben's philosophy. Such an approach does not allow us to formulate any concept of antagonistic subject, which, for instance, could be related to Badiou's subject of fidelity to truth[4] or meet the classical double definition according to which he is, at the same time, an effect of structure and an autonomous agent that actively affects structure.

On the other hand, this ontology of potentiality is connected with the development of a specific type of subjectivity, whose main characteristic is the *passive persistence of bare life*. This feature is most clearly expressed in Agamben's investigation of *homo sacer* and the *Muselmann*, and in the related most problematic construction of the shift from a negative to a positive understanding of these figures, to which he occasionally entrusts his hopes for a "new politics." As is well known, it is precisely these two figures that, already in *Homo Sacer*, represent the main axis around which Agamben develops his theory of sovereign power and the state of exception. The above-mentioned shift is also the central theoretical question of Lorenzo Chiesa's "Giorgio Agamben's Franciscan Ontology,"[5] where the author takes as his starting point the key passage from *Homo Sacer* in which Agamben establishes a structural correlation between the sovereign exception and *homo sacer*:

> Here the structural analogy between the sovereign exception and *sacratio* shows its full sense. At the two extreme limits of the order, the sovereign and *homo sacer* present two symmetrical figures that have the same structure and are correlative: the sovereign is the one with respect to whom all men are potentially *homines sacri*, and *homo sacer* is the one with respect to whom all men act as sovereigns.[6]

As Chiesa stresses, Agamben conceives of the sovereign relation as a *relational relation* between the sovereign exception represented by the figure of the sovereign and bare life as embodied in the figure of *homo sacer*. Chiesa then analyses what he calls the historico-ontological dimension of *homo sacer*, the structure of *sacratio*,[7] drawing attention to the often overlooked fact that Agamben himself explicitly relates the concept of *homo sacer* as defined above to the possibility of a "new politics, which remains largely to be invented."[8] In view of this, it is at first sight surprising that Agamben speaks of a *symmetrical relation* between the sovereign exception and *homo sacer*, but it also seems that the connection between such a symmetrical structure of sovereign power and the possibility of a new politics, whose leading figure is supposed to be precisely *homo sacer*, is by no means coincidental and even confirms this symmetry. The symmetrical relation between, on the one hand, the double exclusion (from the legal jurisdiction and the divine law) that determines the concept of *homo sacer* – and is at the same time "a double capture of his bare life, absolutely exposed to violence and death, in juridical order"[9] – and, on the other, the sovereign exception, can be derived only if we presuppose that *homo sacer* is also the locus where the structure in which he is situated can be transformed. Put differently, Agamben's "new politics" can be thought within the structure of *sacratio* only if we accept that the figure of *homo sacer* is simultaneously the locus where this structure can be suspended. As a consequence, Agamben consistently sees the possibility of a new politics already in the actual extension of the structure of *sacratio* to the entire political space,[10] which he himself conceives as a generalisation of the state of exception that as such "both radicalises – qualitatively and quantitatively – the thanatopolitical expressions of sovereignty . . . and finally unmasks its hidden logic."[11]

It is precisely this line of argument that is confirmed by Agamben's analysis of the modern form of *homo sacer*, the *Muselmann*, which represents the most extreme form of a Nazi concentration camp inhabitant, "the person who in camp jargon was called 'the Muslim', *der Muselmann* – a being from whom humiliation, horror, and fear had so taken away all consciousness and all personality as to make him absolutely apathetic."[12] But Agamben sees a possibility of resistance and suspension of camp power precisely in the fact that "the camp inhabitant was no longer capable of distinguishing between pangs of cold and the ferocity of the SS,"[13] that "he moves in an absolute indistinction of fact and law, of life and juridical rule, and of nature and politics."[14] What Agamben is emphasising here is specifically the *passive persistence of bare life in its indistinction from the laws of the biopolitical structures of sovereignty that produce this very life.*

But we can rightly be sceptical towards this form of political subjectivity. It seems that Agamben's reasoning is theoretically consistent, but I deem it rather difficult to build the concept of a concrete political subject on figures such as Bartleby, *homo sacer* or the *Muselmann.* Agamben possibly relies too much on what we could call the "political effects of passivity."

However, his lapidary recent essay "K.," in which he develops a dialectical discussion of the two characters of K. from Kafka's novels *The Trial* and *The Castle*,[15] cuts deep into this constellation according to which the concept of the subject is torn between the ontology of potentiality embodied in Bartleby and the passive persistence of bare life condensed in the *Muselmann.* If Agamben's critics share the view that an operative theory of the subject should be based on a figure that would be the complete opposite of the camp *Muselmann* who is "absolutely apathetic," an "anti-*Muselmann*" as it were, and if these critics together conclude that such a figure cannot be found in Agamben's work, it is remarkable that, in his interpretation of Kafka's land surveyor K. from *The Castle*, there nevertheless fragmentarily emerges an "anti-*Muselmann*" as an antagonistic subject whose subversive power is not based on the passive persistence of bare life or on a negatively defined ontology of potentiality; rather, land surveyor K. attacks precisely the neuralgic point of the legal order, the form of the law, on which the structure of sovereign power is founded. As I will show, this point is the internal *boundary* that the form of law establishes in the *virtual* state of exception with regard to bare life, so that, despite Agamben's rhetoric, in this state, bare life is never wholly indistinguishable from the law, but exists in a sort of minimal internal difference from the law. It is precisely this boundary that, in the passage from the virtual to the *real* state of exception, the political subjectivity of the land surveyor tries to suspend from within.

If we take into account the fact that Kafka is one of Agamben's constant references, which appears at decisive junctures of his political philosophy, including his elaboration of the concepts of sovereignty and the state of exception in *Homo Sacer*, it is all the more unusual that he hermetically limits the essay in question to the context of Kafka studies without suggesting any further connections to his previous work in the field of politics and philosophy. This is why I will next try to theoretically elaborate on Agamben's interpretation of the protagonists of *The Trial* and *The Castle* and draw all the consequences that follow from it concerning some of the key concepts of his political philosophy. In this regard, we could characterise the circumscribed account advanced in "K." as Agamben's symptomatic transcending of himself, that is, of the line of argument that otherwise strongly dominates his work; we could say that we are here dealing with "Agamben beyond Agamben."

kafkian law

As I have already pointed out, literary figures from Kafka's works often represent an important reference in Agamben's investigations. Kafka's characters appear at crucial passages of his examination of the relation between sovereignty and bare life, a fact which is usually overlooked by critics. It is therefore particularly relevant that Agamben decided to write a separate short essay dedicated to the treatment of some of the key characters of Kafka's novels and short stories. The text focuses especially on the two K.s, the protagonists of *The Trial* and *The Castle*. The filter through which Agamben reads Kafka is, as elsewhere in his work, Roman law. Many interpreters, including his close friend Max Brod, have insisted that the letter K. in the name of *The Trial*'s protagonist "Josef K." actually stands for Kafka himself, thus suggesting that the writer wanted to express half anonymously some of his personal experiences in a fictional story. This line of interpretation has soon become the most influential, also owing to Brod's authority, especially within literary circles. In opposition to this stance, on the basis of the analogy with Roman law, Agamben posits a different thesis, namely, that K. does not mean Kafka, but *kalumnia* and, even more to the point, *kalumniator*.[16]

Kalumniator or *kalumnia* can be translated respectively as a false accuser or the process of a false accusation. Agamben reports that in Roman law this process represented a dangerous threat to the legal system, which can easily be evinced as soon as we dwell on the key implications that the concept of accusation had for law as such.[17] Accusation is, as Agamben claims, the central element of the law; law is based on accusation. Agamben here characteristically bases his argumentation on the etymology of the term and contends that accusation is "perhaps, the juridical 'category' *par excellence* (*kategoria*, in Greek, means accusation), that without which the entire edifice of the law would crumble: the implication of being in the law. The law is in its essence, accusation, 'category'."[18] Slandering is therefore dangerous precisely in so far as, in this process, the very foundations of law are put "on trial" – which possibly explains why in ancient Rome the *kalumniator* was marked for life by branding the letter K on his forehead. But, at the same time, the accusation, which in a way amounts to the trial itself, is the legal *dispositif* through which a subject is himself implicated in the law. For Agamben, belonging to the legal system, the submission to the law is in a strict sense possible only through the process of accusation. In view of his overall interpretation of Kafka's judicial universe, it is therefore even more interesting that he sees in the false accusation through which, according to him, Josef K. triggers the trial against himself[19] also as a strategy whereby the subject tries to subtract himself from the law.

But, as Agamben emphasises, Kafka, for whom "[s]elf-slander is part of a [...] strategy in his incessant struggle with the law,"[20] is well aware of its limitation and insufficiency. We learn this from the other K., the protagonist of *The Castle*, when he says: "[Self-slander] would be a relatively innocent, and in the end quite insufficient, means of defence."[21] Kafka's deliberate question here is, on the one hand, how can self-slander be the means through which one is able to subtract oneself from the pure force of law and, on the other, how can self-slander represent, at the same time, the main element of the procedures by which man is included within the law? According to Agamben, the paradoxical relation between these two strategies (of bare life and law) is most visible in the parable "Before the Law," which is told to K. by the prison chaplain towards the end of *The Trial*.[22] The parable tells of a man from the countryside who reaches the door to the Law. When he intends to gain entry to it, the doorkeeper prevents this from happening. The man from the countryside sits there until his old age, trying to persuade, even bribe the doorkeeper to let him pass. After many years, he finally thinks of the key question: how is it that in all this time nobody else has come by asking to be let in? The doorkeeper's answer is resolute: "Nobody else could have got in this way, as this entrance was meant only for you. Now I'll go and close it."[23] As the prison chaplain explains to K., there is no contradiction between the doorkeeper's first statement forbidding the man from the countryside to enter and his later statement according to which "this entrance was meant only for you"; K. is not being deceived since the prohibition already suggests that the door to the Law is intended for the one whose entry is forbidden. Based on this, Agamben concludes that both statements are an "invitation" to self-accusation. In so far as the law is based on accusation, then the invitation to self-accusation is part of the law's strategy of implicating someone in it.[24] But since self-accusation is at the same time a threat to the very foundations of the law, we should conclude that, as in the structure of *sacratio*, its legal validity wholly coincides with its transgressive function. Thus, in his article "K.," Agamben draws the conclusion that Joseph K.'s whole guilt corresponds precisely to his responding to this invitation to accuse himself by means of self-accusation (self-slander). What distinguishes the fate of Joseph K. from that of the man from the country is that the latter does not respond to such a "call"/invitation and thus subtracts himself from the form of law and its violence.

Having said this, I should stress that Agamben had already established a connection between Kafka's parable and what he calls the form of law in *Homo Sacer*, yet he drew opposite conclusions from it:

> Seen from this perspective, Kafka's legend presents the pure form in which law affirms itself with the greatest force precisely at the point in which it no longer prescribes anything – which is to say, as pure ban. The man from the country is delivered over to the potentiality of law because law demands nothing of him and commands nothing other than its own openness. According to the schema of the sovereign exception, law applies to him in no longer applying, and holds him in its ban in abandoning him outside itself. The open door destined only for him includes him in excluding him and excludes him in including him. And this is precisely the summit and the root of every law.[25]

This, at first sight, clear and unambiguous quotation proves to be much more complex when compared with what Agamben says in "K." In the first place, it seems interesting that, as I said above, in "K.," Agamben comes to a completely opposite conclusion, namely, that the man from the countryside is the one who manages to *subtract* himself from the law as opposed to Josef K., who initiates a self-slandering trial against himself, thus realising the potentiality of the law to be applied to him.[26]

How are we to interpret these apparently contradictory conclusions? It seems to me that we have to pay attention to the fact that Agamben's analysis of the form of law clearly implies two modes of the law: the potentiality of its validity and the realisation of this potentiality. We can therefore suggest that the fates of the man from the countryside and Josef K. differ precisely in their relation to two different modus operandi of the form of law: the man from the countryside manages to keep the form of law in its potentiality, without it being realised, while, with his self-accusation and trial, Josef K. realises this potentiality, the relation of exception characteristic of the law. "Being in force without significance,"[27] which is characteristic of this second modus operandi of the form of law, also explains its key tendency towards the law's complete coincidence with (bare) life. The life of Josef K. and the law become undistinguishable. But with this reasoning we soon run into a new quandary, since, in *Homo Sacer*, Agamben unexpectedly ascribes the life of Josef K. from *The Trial* to the structure of the *virtual* state of exception,[28] the state in which the potentiality of the law to be applied to an exception remains *unrealised*. So how should we understand the confusion created by a comparative reading of *Homo Sacer* and "K."? Are we really facing an elementary contradiction or do we rather need to clarify certain crucial formal differences between the terms and concepts used by Agamben? It seems that this apparent contradiction can be resolved only if we look closely at it from the perspective of the distinction between two types of state of exception, which was already introduced by Walter Benjamin: the virtual and the real. It is only at the basis of this distinction that it is possible to see these two strategies as still caught within the form of law, that is to say, ultimately exposed to the virtual state of exception and thus insufficient to subvert it. As I shall show, this structure can be surpassed only by the passage from the virtual to the real state of exception, in which a third figure emerges, namely, land surveyor K., the subject of subversion. At this stage, let us first elaborate this distinction.

from the virtual to the real state of exception

Agamben develops the difference between the virtual and the real state of exception starting from Benjamin's eighth thesis from "Theses on the Philosophy of History":

> The tradition of the oppressed teaches us that the "state of emergency" in which we live is not the exception but the rule. We must attain to a conception of history that is in keeping with this insight. Then we shall clearly realize that it is our task to bring about a real state of emergency . . .[29]

From these remarkable lines, which represent Benjamin's reaction to the Hitler–Stalin Pact, in *Homo Sacer*, Agamben infers the following:

> We have seen the sense in which law begins to coincide with life once it has become the pure form of law, law's mere being in force without significance. But insofar as law is maintained as pure form in a state of virtual exception, it lets bare life (K.'s life, or the life lived in the village at the foot of the castle) subsist before it. Law that becomes indistinguishable from life in a real state of exception is confronted by life that, in a symmetrical but inverse gesture, is entirely transformed into law. The absolute intelligibility of a life wholly resolved into writing corresponds to the impenetrability of a writing that, having become indecipherable, now appears as life. Only at this point do the two terms distinguished and kept united by the relation of ban (bare life and the form of law) abolish each other and enter into a new dimension.[30]

We can now understand in what sense the fate of the man from the countryside and that of Josef K. from *The Trial* (as well as that of the villagers who live at the foot of the castle) can all be considered at the level of the virtual state of exception. The difference between the virtual and the real state of exception concerns processes that are not simply symmetrical but opposite; it is a qualitative difference. In the virtual state of exception, there still persists the internal boundary separating law from life, which the law wants to coincide with. The law is still confronted by life, albeit bare life. In this sense, we can agree with Agamben that the relation between the sovereign exception and bare life is symmetrical in so far as the law that is in force without significance is confronted by bare life, life that exists without signifying, that is, pure nothingness.[31] But if we look at this relation from the other, opposite side (that is, if we follow Benjamin) we see that it is only life that wants to pass entirely into law that produces the key qualitative change. Put differently, "[c]onfronted with the imperfect nihilism that would let the Nothing subsist indefinitely in the form of a being in force without significance, Benjamin proposes a messianic nihilism that nullifies even the Nothing and lets no form of law remain in force beyond its own content."[32]

Such a definition of the difference between the two types of the state of exception, the virtual and the real, provides us with an answer to the question as to how or with which operation Benjamin achieves the qualitative transformation of the relation of ban existing between the form of law and bare life. If it holds that, in the virtual state of exception, there still persists an internal boundary between the law and bare life, or nothing, which the law wants to coincide with, then we cannot derive the concept of the real state of exception only by changing our point of view, that is, by looking at the relation from the standpoint of the second term; there must be something else, an operation that formally transcends the existing structure of the relation between the form of law and bare life. It is not simply a matter of an inversion in which *homo sacer* merely takes the place of the sovereign exception and thus produces its own subjectification. From this perspective, we can rightly be sceptical when Agamben speaks of a "symmetrical but inverse gesture."[33] If it holds that, in the real state of exception, both terms (the form of law and bare life) "abolish each other and enter into a new dimension," then this is possible only because bare life, which is transformed into law, carries out an *operation of deactivation* by which it *cancels* or *suspends the boundary* separating it from the law, the boundary that is the key element of the structure of sovereign exception or the virtual state of exception. But, thus, this operation also suspends the law itself. What is crucial in Agamben's short contribution on Kafka is that he unknowingly produces the figure of the subject of this operation: land surveyor K. from *The Castle*.

joseph k. and land surveyor k.

Agamben divides his concise text "K." into two parts. In the first, he focuses on the interpretation of the figure of Josef K. from *The Trial*; this part also includes, in addition to a discussion of Roman law, numerous references to Kafka's personal correspondence and diary entries through which key passages and issues of the novel are given a detailed interpretation. In the second part, Agamben analyses the figure of the other K., the land surveyor from *The Castle*. This schematic division already suggests that, in his opinion, there is a complex relation between the two protagonists. If we accept Agamben's main thesis that *The Trial* is centred on the case of Josef K.'s self-accusation, with which he tries to elude the grip of the law, then we cannot overlook the importance of the fact that this strategy is ultimately ineffective, as evidenced by the dismal fate of Josef K., who is killed. At the same time, the impasse of such a strategy becomes even more obvious if we compare it with the truly subversive operation carried out by the figure of land surveyor K. from *The Castle*.

This is why I shall next try to verify the thesis according to which Benjamin's real state of exception is not merely a "symmetrical but inverse" relation to that which exists between the form of law and bare life as they appear in the virtual state of exception; rather, the shift from the virtual to the real state of exception is to a great extent determined by the specific action of figures such as land surveyor K. (or even land surveyors in general as present in Roman law), one that erases or suspends the boundary between law and bare life, that is, the key element of the relation between these two terms as it exists in the virtual state of exception. Put differently, the figures of both K.s share the same goal, namely – to use Agamben's jargon – deactivating and rendering inoperative the law, but within this framework it is only land surveyor K. who manages to aim at the truly neuralgic point of the law – the boundary that the law establishes between itself (the castle) and bare life (the village beneath the castle).

As I have already remarked, Agamben sets his interpretation of Kafka in the context of Roman law, which the latter studied intensively in his training for the legal profession. Following Agamben's analysis, the central point of *The Trial* is the strategy of self-accusation with which Josef K. tries to subtract himself from the law. But already the fact that (self-)accusation – which, in view of its etymology and the historical sources that Agamben refers to, is the central constitutive element of law in general – is Josef K.'s only strategy implies that, at the structural level, it corresponds to what, in *Homo Sacer*, Agamben calls life that is "wholly resolved into writing." Josef K.'s self-accusation thus presupposes precisely the attempt of making his life pass into the law and, with this move, gradually blurring the difference between the two. Already in *Homo Sacer* Agamben concludes that "the existence and the very body of Joseph K. ultimately coincide with the Trial; they *become* the Trial."[34]

This kind of relation between life and law is also connected with another of Kafka's central motifs: torture. The intensity of Kafka's obsession with the idea of torture is striking and can be inferred from numerous notes in his diaries and personal letters, while, with regard to his fictional works, we find clear evidence of this in the short story *In the Penal Colony*. But we have to be careful about the precise function of this theme. Following Agamben's interpretation, for Kafka, torture was more a means of legally extracting a confession than a goal in itself. In Kafka's work, torture occupies the same place as it did in Roman law's search for truth. And, as Agamben argues, in Roman law, the primary purpose of such means was precisely the extraction of the self-condemnation of the person under investigation.

If we now refer to the distinction Benjamin makes, in "Critique of Violence,"[35] between the mythical violence of the law (law-making and law-preserving violence) and divine violence, which destroys the law, we can see that Kafka's understanding of torture is to be located on the plane of mythical violence, i.e., violence as a means of preserving the law. This is why we can justifiably ask whether, in *The Trial*, Kafka defends the law rather than deactivating it. Here, torture that leads to self-condemnation appears as the extralegal means, in short, the violence, that the law activates precisely at the moment when it feels most threatened. But such a conclusion also overlooks the fundamental ambiguity traversing *The Trial*. On the one hand, the motif of torture does suggest that Kafka advocates the stand of the mythical law-preserving violence, which defends the law, whereby he also advocates the stand of defending tradition, which is for Benjamin always barbaric. But, on the other hand, Kafka certainly wants to point to the unsuccessfulness of the strategy of self-slander and consequently to the unsuccessfulness of the attempt to abolish the law by way of passing into it. In this sense, Josef K.'s fundamental guilt is self-slander, which is in a way equivalent to the guilt of bare life as the object of sovereign violence. In short, both conclusions eventually point to the inappropriateness or the ineffectiveness of self-slander as a method of struggle against the law.

land surveyor k.: agamben's anti-*muselmann*

The truly dialectical moment in Kafka emerges in *The Castle*. In this novel, the writer explicitly rejects the strategy of self-slander as a "relatively innocent, and in the end also quite insufficient, means of defence."[36] The fundamental question is then the following: wherein lies the subversiveness of land surveyor K. in comparison with the defeat of Josef K. from *The Trial*? Agamben already discusses the land surveyor – *agrimensor* in Latin – and his activity as detailed by Roman law in *Homo Sacer*; here he identifies him with those actions for which individuals were ascribed the status of *homo sacer* par excellence:

> The crimes that, according to the original sources, merit *sacratio* (such as *terminum exarare*, the cancellation of borders; *verberatio parentis*, the violence of the son against

> the parent; or the swindling of a client by a counsel) do not, therefore, have the character of a transgression of a rule that is then followed by the appropriate sanction. They constitute instead the originary exception in which human life is included in the political order in being exposed to an unconditional capacity to be killed. Not the act of tracing boundaries, but their cancellation or negation is the constitutive act of the city (and this is what the myth of the foundation of Rome, after all, teaches with perfect clarity).[37]

We see how, in *Homo Sacer*, Agamben does not yet thematise the difference between the sovereign exception and *homo sacer*, but thinks them in their structural symmetry and even in their coincidence. The erasure and negation of boundaries, which is what a land surveyor ultimately does, is here thought in a strict correlation with the cancellation of borders that is constitutive of a city and lies in the exclusive domain of the sovereign exception. The land surveyor's status of *homo sacer* testifies to the sacred status ascribed to boundaries in Rome: "Indeed, the border of Rome had a sacred character to such an extent that he who erased the borders (*terminum exarare*), became *sacer* and could be killed with impunity by anyone."[38] That the erasure of borders is the land surveyor's sole interest becomes even clearer precisely in Kafka's *The Castle*. Agamben's interpretation here shows a perfect agreement between the land surveyor of Roman law and the figure of Kafka's land surveyor K. The latter, too, aims at erasing borders, that is, those separating and, at the same time, keeping bound together the castle and the village:

> Rather, since life in the village is, in reality, entirely determined by the borders that separate it from the castle and at the same time keep it bound to it, it is above all these boundaries that the arrival of the land surveyor calls into question. The "assault on the last frontier" is an assault against the boundaries that separate the castle (high) from the village (low).[39]

This crucial excerpt clearly explains wherein lies the land surveyor's *differentia specifica* with regard to the figure of the *Muselmann*. The structural relation between the castle and the village is the relation that determines the virtual state of exception in which, as I have already pointed out, the key factor is the *internal boundary* that inherently separates the law from bare life as pure nothing by keeping them at a sort of internal distance. The Latin *terminus technicus* that denotes this boundary is *kardo*, the line from the North to the South poles, which is named as such since "it directs itself towards the cardinal point of the sky"[40] and on which basis land surveyors establish borders. But *kardo* also refers to the hinge of the door. Here, Agamben refers to Isidore's etymology of this term:

> A hinge is the place in which the door (*ostium*) turns and is moved, and is so called from the Greek term for heart (*apo tes kardias*), for just as the heart of man governs every thing, so the hinge rules and moves the door. Hence the proverb: *in cardinem esse*, to be at a turning point.[41]

And the singularity of the figure of the land surveyor is to be found precisely in the fact that he is the only one who, due to his ability to erase boundaries, attacks these internal boundaries and can actually cancel them. He is the one who finds himself "at a turning point," at the extreme internal edge of the structure of *sacratio* on which it is possible to suspend sovereign power and the virtual state of exception from within. But there is no structural symmetry between the sovereign power of the castle and the activity of the land surveyor. The aim of Kafka's land surveyor is not the establishment of new boundaries or the shifting of old ones; the arrival of the land surveyor in the village is a pure sign of the abolition of boundaries as such, the boundaries that establish the form of law in the virtual state of exception. As Agamben says:

> And if Bucephalus [a character from Kafka's 1919 story *Der neue Advokat*] is the "new lawyer", who studies the law only on the condition that it is no longer applied, K. is the "new land surveyor" who renders inoperative the boundaries and borders that separate (and keep bound together) the high and the low, the castle and the village, the temple and the house, the divine and the human.[42]

It is telling that land surveyor K. comes to the village without his land surveying instruments, carrying only a "knobbly walking stick" with which he tries to destroy the boundaries of the sovereign power of the castle.[43]

Here it is important to add that it is precisely this internal boundary, or *kardo*, that marks the intermediary area of half-human figures, so significant for Kafka's work: doormen, secretaries, lawyers, who represent the law. The crucial role that these figures occupy shows us how, for Kafka, in the last instance, the whole problem of the law emerges as the problem of the existence of its mediators. What is dangerous for Kafka is the fact that

> Even the priest is a doorkeeper of the law, even he "belongs to the court"; and the true deception is, precisely, the existence of doorkeepers, of men (or of angels: in the Jewish tradition, guarding the door is one of the functions of the angels), extending from the lowest functionary up to the advocates and the *highest judge*, whose purpose is to induce other men to accuse themselves, to make them pass through the door that leads nowhere except to the trial.[44]

These worldly representatives of the law embodied in the figures of guards, lawyers and judges are those who, for Kafka, represent an intermediary half-human world between nature and law. As Eric Santner has noted,

> such figures, partaking in characteristics of animal and machine, occupy an uncanny ontological domain – call it the *subaltern sublime*. They serve to mediate the transfer, the conversion of the materiality of meaningless, physical causes – the rote repetition of dead letters – into ideal, symbolic effects. They inhabit a space comparable with that impossible frontier that Freud identified as the locus of the drives, the site where blind nature begins to exceed itself – where merely *sinnlich* becomes *über-sinnlich* – and is thereby converted into culture.[45]

From Kafka's perspective, this ghastly world of half-human images is the boundary that separates the law from its outside. It represents the internal boundary of law itself, the main constitutive condition of the operation of the "form of law," a law that is emptied of all significance so that what remains is the bare force of its validity, a "being in force without significance."

It is therefore not surprising that the arrival of the land surveyor poses a threat primarily to such characters and by no means directly to the sovereign power of the castle represented by the absent Count Westwest. As Agamben's biographical remarks on Kafka explain, these characters represent the boundaries that governed the society of Western Jewish intellectuals, which the writer belonged to. Thus, they are all an equivalent to the doors in the parable "Before the Law," to their function of closing life into the law and, at the same time, separating the two internally. The hinge of the door as the "turning point" is "the place in which the door, which obstructs access, comes to be neutralised."[46] In this context, writing represents the weapon with which Kafka tries to suspend the boundaries of the social world he was part of.[47]

from the violence of boundaries to the violence against boundaries

Now, we only need to explain the land surveyor's gesture of erasing the boundaries in light of its coincidence with Benjamin's notion of divine violence. This step is important since Agamben himself says that this violence stands "in the same relation to sovereign violence as the state of actual exception, in the eighth thesis, does to the state of virtual exception."[48] I have already pointed out the qualitative heterogeneity of the actual state of exception in relation to the virtual: the same heterogeneous relation holds between divine and sovereign violence. At this point, we should refer to the last pages of Benjamin's "Critique of Violence," where he explains the key difference separating the mythical, law-making violence from divine violence:

> Far from inaugurating a purer sphere, the mythic manifestation of immediate violence shows itself fundamentally identical with all legal violence, and turns suspicion concerning the latter into certainty of the perniciousness of its historical function, the destruction of which thus becomes obligatory. This very task

> of destruction poses again, in the last resort, the question of a pure immediate violence that might be able to call a halt to mythic violence. Just as in all spheres God opposes myth, mythic violence is confronted by the divine. And the latter constitutes its antithesis in all respects. *If mythic violence is lawmaking, divine violence is law-destroying; if the former sets boundaries, the latter boundlessly destroys them;* if mythic violence brings at once guilt and retribution, divine power only expiates; if the former threatens, the latter strikes; if the former is bloody, the latter is lethal without spilling blood [...] On the breaking of this cycle maintained by mythic forms of law, on the *suspension of law with all the forces* on which it depends as they depend on it, finally therefore on the abolition of state power, a new historical epoch is founded [...] Divine violence, which is the *sign and seal but never the means* of sacred execution, may be called sovereign [*waltende*] violence.[49]

What we have already observed in our analysis of Agamben's interpretation of the "form of law" becomes even clearer here. Benjamin says specifically that the mythical, law-making violence is the one that "sets the boundaries" while divine violence is committed to their destruction. And since Benjamin's essay is set explicitly in the context of law, we can justifiably say that the boundary that in the form of law separates law from bare life, which it confronts, and the boundary set by the law-making, mythical violence are one and the same thing. Conversely, this is why the sovereign erasure of boundaries that Agamben identifies as the constitutive act of the city and the land surveyor's negation and erasure of boundaries are qualitatively different. The first belongs to the order of mythical violence and is a means of establishing the law, the form of law, the law that is in force without signifying and, last but not least, the virtual state of exception; on the other hand, the land surveyor's erasure and negation of boundaries is a mere sign that manifests a tendency to abolish state violence and its boundaries. Put differently, as Žižek has it:

> the opposition of mythic and divine violence is that between the means and the sign, that is, mythic violence is a means to establish the rule of Law (the legal social order), while divine violence serves no means, not even that of punishing the culprits and thus re-establishing the equilibrium of justice. It is just the sign of the injustice of the world, of the world being ethically "out of joint."[50]

In so far as divine violence is qualitatively heterogeneous from other kinds of violence, in so far as it "is not one kind of violence among others, but only the dissolution of the link between violence and law," Benjamin can say that "divine violence neither posits nor conserves violence, but deposes it."[51] From this perspective, we can draw our conclusion: the actual state of exception from Benjamin's eighth thesis is equivalent to divine violence from his "Critique of Violence." In this sense, we can say that the Kafkian figure of the land surveyor is the subject of both divine violence and the production of the actual state of exception – Benjamin's legacy to the historical materialist. It is precisely in this constellation that we can suggest that land surveyor K. represents the kind of subjectivity that condenses the subversive power aimed at the abolishment of the structures of sovereign power and the virtual state of exception, a subjectivity that cannot be found anywhere else in Agamben's work, but can be derived from the key concepts of his philosophy.

notes

1 See Giorgio Agamben, *Homo Sacer: Sovereign Power and Bare Life*, trans. Daniel Heller-Roazen (Stanford: Stanford UP, 1998) 11.

2 For this line of criticism, see Nina Power, "Potentiality or Capacity? Agamben's Missing Subjects," *Theory & Event* 13.1 (2010), available <http://www.mediafire.com/?g5nzkddywz2>.

3 In his essay on Bartleby, Agamben says the following:

> Since Aristotle stated that all potentiality is also potentiality not (to be or do), the theologians were forced to strip God of all potential to be and to will at the same time

> that they affirmed his omnipotence. If God had the potential to be, he could also not be, which would contradict his eternity. On the other hand, if God were capable of not wanting what he wants, he would be capable of wanting non-Being and evil, which is equivalent to introducing a principle of nihilism into God. (Giorgio Agamben, *Potentialities: Collected Essays in Philosophy*, trans. and ed. Daniel Heller-Roazen (Stanford: Stanford UP, 1999) 253)

For Agamben, all ethical traditions of Western metaphysics "sought to avoid the problem of potentiality by reducing it to the terms of will and necessity. Not what you *can* do, but what you *want* to do or *must* do is [their] dominant theme" (ibid. 254).

4 See, for example, Alain Badiou, *Being and Event*, trans. Oliver Feltham (London and New York: Continuum, 2006).

5 See Lorenzo Chiesa, "Giorgio Agamben's Franciscan Ontology" in *The Italian Difference*, eds. Lorenzo Chiesa and Alberto Toscano (Melbourne: re.press, 2009) 149–63.

6 Agamben, *Homo Sacer* 84.

7 See Chiesa 149–56.

8 Agamben, *Homo Sacer* 11.

9 Chiesa 149.

10 "The decisive fact [is] that, together with the process by which exception everywhere becomes the rule, the realm of bare life – which is originally situated at the margins of the political order – gradually begins to coincide with the political realm" (Agamben, *Homo Sacer* 9).

11 Chiesa 151–52.

12 Agamben, *Homo Sacer* 185.

13 Ibid.

14 Ibid.

15 See Giorgio Agamben, "K.," trans. Nicholas Heron, in *The Work of Giorgio Agamben: Law, Literature, Life*, eds. Justin Clemens, Nicholas Heron, and Alex Murray (Edinburgh: Edinburgh UP, 2008) 13–27.

16 At the very beginning of his essay, Agamben reminds us that, while preparing for the legal profession, Kafka studied the history of the Roman law. See Agamben, "K." 13.

17 Ibid.

18 Ibid. 15.

19 The first sentence of *The Trial* reads as follows: "Someone must have slandered [*verleumdet haben*] Josef K., he knew he had done nothing wrong but, one morning, he was arrested" (Franz Kafka, *The Trial*, trans. David Wyllie (Teddington: The Echo Library, 2007) 1; translation modified). Based on an analogy with the figure of the *kalumniator*, Agamben infers that this "someone" who slandered Josef K. is Josef K. himself. See Agamben, "K." 13. In the English translation of Agamben's "K.," the key term – the Latin *kalumnia* (or *kalumniator*), the Italian *calunnia* (or *calunniatore*), which are equivalent to the German *Verleumdung* (or *verleumden*) used by Kafka – is appropriately translated as *slander*, which has the same juridical meaning as *kalumnia*.

20 Agamben, "K." 15.

21 Franz Kafka, *The Castle*, trans. Mark Harman (New York: Schocken, 1998) 252.

22 Idem, *The Trial* 128–29.

23 Ibid. 129.

24 This "invitation to self-accusation" (we may refer to it as a "zero degree of interpellation") is what Slavoj Žižek calls "interpellation without identification," which, according to him, is one of the central themes of Kafka's novels:

> In a first approach, the starting point in Kafka's novels is that of an interpellation: the Kafkaesque subject is interpellated by a mysterious bureaucratic entity (Law, Castle). But this interpellation has a somewhat strange look: it is, so to say, an *interpellation without identification/subjectivation*; it does not offer us a Cause with which to identify – the Kafkaesque subject is the subject desperately seeking a trait with which to identify, he does not understand the meaning of the call of the Other. (Slavoj Žižek, *The Sublime Object of Ideology* (London and New York: Verso, 1989) 44)

25 Agamben, *Homo Sacer* 49–50.

26 Idem, "K." 21. It has to be said that this possibility is already hinted at in *Homo Sacer* – see idem, *Homo Sacer* 55ff.

27 This phrase is derived from the correspondence between Benjamin and Scholem in which they exchanged their views precisely on Kafka's parable – see Agamben, *Homo Sacer* 51.

28 Agamben, *Homo Sacer* 55.

29 Walter Benjamin, "Theses on the Philosophy of History" in his *Illuminations*, ed. Hannah Arendt (New York: Schocken, 1969) 253–64 (257).

30 Agamben, *Homo Sacer* 55.

31 This is what Scholem advocates in his polemics with Benjamin.

32 Agamben, *Homo Sacer* 53.

33 It seems that, in *Homo Sacer*, precisely in the case of the Roman land surveyor, who represents the figure of *homo sacer* par excellence and whose main characteristic is – like in the case of the sovereign – "the erasing of boundaries," Agamben insists on such a conclusion, which is problematic (see *Homo Sacer* 85). At this level, we should then talk about two sovereign exceptions: the exception of *homo sacer* according to the sovereign and the exception of the sovereign according to *homo sacer*. But the difficulty of such a conclusion lies in the fact that we thus lose the difference between the virtual and the real state of exception, that is, one of Agamben's key presuppositions.

34 Agamben, *Homo Sacer* 53.

35 See Walter Benjamin, "Critique of Violence" in his *Reflections*, ed. Peter Demetz (New York: Schocken, 1986) 277–300.

36 Kafka, *The Castle* 252.

37 Agamben, *Homo Sacer* 85.

38 Idem, "K." 25. Here, the basic ambivalence of the concept *sacer* that Agamben repeatedly talks about is finally clarified; it does not amount to the double meaning of the term, which is Benveniste's thesis; we also must not place all the emphasis on the issue of double exclusion, which is what Agamben does in *Homo Sacer* (see 81–86) (perhaps this is one of the reasons why he finally slips into a vitalistic conception of the form of life . . .). As Jean-Claude Milner contends in his unsurpassed analysis, it is rather decisive "that *sacer* [. . .] means neither sacred nor damned, but the relation between both terms: namely, the dimension of the boundary between the pure and the impure" (Jean-Claude Milner, *Le Périple structural. Figures et paradigme* (Paris: Seuil, 2002) 72). This reading is possible only if, instead of the double meaning of *sacer*, we introduce the distinction between meaning and referent: "The *referent* of the word *sacer* consists of two opposite attitudes at the core of one and the same culture, but it has only one *meaning*. It seems that this meaning will be defined as the pure position of a boundary" (ibid. 73). As we shall see, the subversiveness of the land surveyor proceeds precisely from his indistinction according to the boundary set by the form of the law: the land surveyor is the one who finds himself at a "turning point," he is the *relation*.

39 Agamben, "K." 25.

40 Ibid. 23; translation modified.

41 The citation is taken from *Etymologiae* (or *Origines*) by Isidore of Seville, a compilation of ancient etymologies from the seventh century (see Agamben, "K." 26).

42 Agamben, "K." 26.

43 Ibid. 24.

44 Ibid. 21.

45 Eric L. Santner, *My Own Private Germany: Daniel Paul Schreber's Secret History of Modernity* (Princeton: Princeton UP, 1996) 75–76.

46 Agamben, "K." 26.

47 On 16 January 1922, Kafka wrote in his diary (cited from Agamben, "K." 24–25):

> All such writing is an assault on the frontiers; if Zionism had not intervened, it might easily have developed into a new secret doctrine, a Kabbalah (*zu einer neuen Geheimlehre, einer Kabbala*). There are intimations of this. Though of course it would require a genius of an unimaginable kind to strike root again in the old centuries, or create the old centuries anew, and not thereby expend its forces, but begin only now to consummate them.

48 Agamben, *Homo Sacer* 65.

49 Benjamin, "Critique of Violence" 249–52; my emphasis. A terminological clarification is much needed here: what Benjamin finally

names "sovereign violence" (*waltende Gewalt*) or divine violence is the opposite of Agamben's conception of the violence exercised by the sovereign exception. The latter is, for him, following Carl Schmitt, always a law-making exception, and consequently it corresponds to Benjamin's mythical (i.e., non-sovereign/ non-divine) violence.

50 Slavoj Žižek, *Violence* (New York: Picador, 2008) 199–200.

51 Agamben, *Homo Sacer* 65.

lorenzo chiesa
frank ruda

THE EVENT OF LANGUAGE AS FORCE OF LIFE

agamben's linguistic vitalism

The aim of this paper is threefold. Firstly, we intend to emphasise the systematic nature of Agamben's project from his early work to his most recent publications, a project that insistently proposes a supposedly new, but in the end quite traditional, definition of philosophy. Secondly, we mean to show how such an endeavour is first and foremost ontological, not political, and explicitly inscribes itself within the legacy of twentieth-century philosophy's (especially Heidegger's) attempt to overcome metaphysics. Thirdly, we seek to problematise the proximity, all too often taken for granted, between Agamben's ontological politicisation of philosophy and Badiou's and Žižek's re-launching of a "communist hypothesis" that is inextricable from a positive re-evaluation of materialism and dialectics. In a few words, our claim is that Agamben is a *sui generis vitalist* thinker and his recuperation of dialectics (what he prefers to call "the bipolar machine")[1] can only be understood in this framework, that is, outside, if not against, any return to Marx.

We also advance the proposition that the second and third points above should be read together: Agamben manages to distinguish his project from Heidegger's (a constant programme throughout all his work, from *Language and Death* (1982) to *The Kingdom and the Glory* (2007) passing through *Homo Sacer* (1995), *The Time that Remains* (2000), and *The Open* (2002)) only at the price of promoting a sophisticated type of linguistic vitalism, which can be encapsulated by the phrase "form of life." Moreover, such a linguistic vitalism ultimately radicalises Heidegger's own ontological discourse without refuting it; in other words, in our opinion, in trying to overcome Heideggerian thought, Agamben actually refines it and brings it to its most extreme consequences. It is now a matter of seeing how this happens.

In brief, our central argument is the following: for Agamben, life is a negativity, or potentiality, *independently of history itself*. This is for us the ultimate meaning of the notion of "form of life." That is to say, nature is as such somehow "historical" in a different sense independently of history understood as Heidegger's forgetting of the ontological difference. At the level of what we could call historical history, that of the forgetting of the ontological difference, Agamben introduces a critique of politics: he develops politically Heidegger's reflection on the epochality of being/nihilism in terms of the dialectic between sovereignty and exception.[2]

But the negativity/potentiality – or historicity – of the form of life as such cannot be confined to this level, that which Heidegger would call "metaphysical." Agamben's confrontation with Heidegger is somehow paradoxical in so far as he manages to challenge the latter's deconstruction of metaphysics only by opening up a new dimension of negativity/potentiality that is understood as affirmatively *meta-metaphysical*, meta-historically historical.[3]

More generally, we believe that Agamben's ontological politicisation of philosophy intends to collapse what he calls the "form of law" into an undetermined form of *life*: note that here "form of law" means quite simply the *metaphysical* form of life, in Heidegger's parlance, the form of life which characterises the epoch of nihilism and the forgetting of the ontological difference (this coincidence is also proved by Agamben's recurrent assumption of the existence of something as monolithic as "the substance of the Occident," "the tradition of Western philosophy," "Western humanity").[4] Also, this *living* process of indetermination of the immanent relation between life and form (i.e., ultimately, that which constitutes the speaking being) can remain undetermined only if it continuously repeats itself. In other words, we can overcome the form of law as metaphysical form of life only by re-enacting what in more general philosophical terms we could call the *genesis of the transcendental* (the "as not" that defines the peculiarity – the formal in-determination – of the living species *Homo sapiens*).

It is our contention that this ontological discussion can easily be applied to the two most ambitious developments of Agamben's politics: the coming community and the Messianic community – or better, we are tempted to say, the community to come as redeemed Christianity... The problem is that because of these very ontological premises, Agamben is unable to delineate any positive notion of political agency. In Hegelo-Marxian terms, Agamben's political project cannot promote a political subject because it replaces the continuous re-determination of in-determination – the latter is, for Agamben himself, both ontological and biological; for instance, he speaks respectively of an "essential inoperativity" and of the "Sabbatical animal"[5] – with the *repetitive reification of* pure *indetermination* (or potentiality); all of Agamben's well-known political examples, or "paradigms," from Bartleby to Tiananmen and the *Muselmann*, could be read in this way. In other words, Agambenian politics does not give rise to any positive political subject in so far as it relies on an ontological subsumption of *re*-determination under in-*determination.*[6] Indetermination thus becomes a *ur*-determination and Agamben's ontological politicisation of philosophy presupposes a substantialist stance.[7]

This same argument could also be reformulated with regard to the notion – dear to Agamben – of potentiality. As we will explain in detail below, Agamben implicitly presupposes *two* kinds of potentiality, hierarchically related: the potentiality *not to* and the *potentiality* not to, where the former is dependent on the latter; that is, reified in-*determination* ultimately determines negativity. From a slightly different perspective, returning to our initial point on nature and history, this means that historical (political) potentiality is inscribed in nature as such: this is what we defined as Agamben's vitalism *sui generis*. The "event of language" – a phrase he uses in crucial passages of his work, from *Language and Death* to *The Sacrament of Language* – the emergence of history *sensu stricto* (which, as we have seen, Agamben reads through a Heideggerian lens) is logically preceded by the genesis of the transcendental as immanent to nature (the *potentiality* not to). This fundamental distinction locates Agamben's project among what has aptly been described as philosophies for which "every form-of-life reduces itself to the transitory and precarious expression of a force-of-life" and "there is being only because there is life."[8] Agamben's ultimate ontological aim is an understanding of the "*nature* of thought" from the perspective of "life [...] as a power that incessantly exceeds its forms and realisations."[9] Speaking about Deleuze's philosophy, he distinguishes two kinds of vitalism: for the first there is act without essence, for the second there is potentiality (*potenza*) without action.[10] In what follows, we will investigate the way in which, in the wake of the French philosopher, Agamben opts for the latter.

•••

As Agamben put it in a discussion with Alain Badiou on the occasion of the publication of the French translation of *Coming Community*,

> we have to think the event. But can we think an event of language [*langage*] itself? Not of speech [*langue*], but the taking place of anything as the being-called of language [*l'être dit du langage*]; if this is possible, an event such as the coming of the community could have a meaning.[11]

Agamben clearly explains why it is necessary to think an event of language itself: it is because any "antinomy of the individual and the universal has its origin in language."[12] More precisely, this means that thinking the event of language itself ultimately allows us to elaborate a new – non-antinomical – way of addressing the question of the relationship between the individual and the universal, which would neither posit the individual as a particular embodiment of the universal nor conceive of the universal as being present only in individual embodiments. For Agamben, these two options encapsulate the specific shortcoming of dialectics: the presupposition that there is a dialectical relationship between the individual or particular and the universal misses the fact that the particular is always *more* particular than any particular embodiment of the universal, that is, it exists as a singularity. In turn, one equally fails to think the universal if one conceives it on the basis of its relation with a particular that is not singular.

Here Agamben's approach, as he promptly admits, runs counter to the entire history of Western metaphysics: the latter is in fact caught in this dialectical misunderstanding, which is best exemplified by its interpretation of the relation between potentiality and actuality in Aristotle. If any actuality is simply read as a particular realisation of a more universal potentiality, then the series of all individual acts, the totalised unity of all particular actualities, would be nothing but the fully realised universal potentiality which logically preceded them and from which they originated. The universal, potentiality, would coincide with the condition of possibility of actuality, the particular, and thus amount to a transcendental. Against this version of the metaphysical tradition, Agamben claims that

> the whole problem is knowing how one conceives of the transcendental [...] One maybe needs to contract possibility and necessity (as the late Schelling did) [...] one needs to think whatever being [*le quelconque*] that is not indifference. And this is the transcendental.[13]

Agamben's project therefore addresses a transcendental question in order to rethink the relationship between the universal and the particular; to achieve this purpose, he needs to counter most of Western metaphysics and eventually develop a theory which makes it possible to conceive of *whatever* is singularly as a *being* whatever. It should be evident by now that such an enterprise has a straightforwardly ontological focus – for Western thought, the theory that can render intelligible whatever is and can be is ontology; it addresses the question of being which needs to be re-articulated by working through the connection between the universal and the particular.

From a slightly different perspective, Agamben's general project could also be summarised as follows: he aims at establishing a theory of the event in and of language according to which being as such is meta-metaphysically a sort of arche-event. How should we understand this polemical claim? In "What is a Paradigm?," one of the three long essays contained in *The Signature of All Things* (2008), Agamben refers to everything he has developed up to that point in his philosophy – from the elaboration of the notion of *homo sacer* to the theory of the camp, from the relevance of the figure of the *Muselmann* to the economic interpretation of Trinitarian theology – as a "paradigm,"[14] and he claims that any archaeologist – and, as is well known, he considers himself to be one – also has to be a paradigmatologist. Relating this bold statement to his discussion with Badiou, we can conclude that, for Agamben, the evental character of language – which concerns whatever being – can only be grasped properly by means of paradigms. But, more specifically, how does he determine what a paradigm is?

It is "a singular object that, standing equally for all others of the same class, defines the intelligibility of the group of which it is a part and which, at the same time, it constitutes."[15] A paradigm is first and foremost a singularity, a singular object. But to become or be a singularity the paradigm has to be subtracted from its ordinary context and use in order to be able to present what we could call the "rule" of its ordinary use.[16] A paradigm can only stand for all objects of its class when, at the same time, it is excluded from this very class. To put it simply, a paradigm – for example, of a performative speech act – is excluded from the class of objects it stands for. The example of a performative speech act is itself not a performative speech act; saying that "I marry you" is an example of a performative speech act that does not entail marrying someone. Therefore, the rule a paradigm presents or, more precisely, the rule it constitutes – the example of a performative speech act presents the way in which one ordinarily uses a performative speech act, how it usually functions – can neither be a universal rule that a priori grounds all particular cases and would thus be deducible, nor can it be rendered intelligible by gathering all particular cases – of performative speech acts, for example – and therefore be gained inductively. When it comes to paradigms there is always a different relationship between particularity and universality involved. This is the case because the condition of possibility of a paradigmatic relation in language is the exclusion of the paradigm from the set it stands for and exemplifies. As Agamben has it,

> the proper place of the example is always beside itself, in the empty space in which its undefinable and unforgettable life unfolds. This is purely linguistic life [. . .] Exemplary being is purely linguistic being. Exemplary is what is not defined by any property, except being-called.[17]

Thereby paradigmaticity is the retroactive effect of the necessary exclusion of a singular element – i.e., of an element of language, of an example – that enables, or, put another way, constitutes the dimension of universality as such – the latter can thus no longer be referred to in this way. In other words, not only does a singularity stand for more than just an element that, as we have seen, cannot be confined to the particular, but also directly produces "universality," for instance, that of all performative speech acts. The paradigm thus (paradigmatically) presents a "potentiality not to" be inscribed into language which is condensed in the possibility of giving an example – a specifically human capacity – that is, precisely, the potentiality of the "not to"[18] which the example is. This is why Agamben insists on the fact that any paradigm is the coincidence of a singularity and its "pure exposure,"[19] or, to return to our previous discussion, what one may call an event in language. Paradigms are events in and of language that present the pure being-called of whatever being (i.e., becoming an example).[20]

But to properly grasp what is at stake we also have to take into account how different examples or paradigms relate to each other, even within Agamben's own philosophy. This is, at the same time, precisely a way of understanding why, in the case of examples or paradigms, no definition and no property is involved. When Agamben claims that they can be conceived of as being in a relation neither of particularity to universality nor of universality to particularity, he is relying on the notion of the "coming community." More specifically, he tells us that the relationship between examples or paradigms can be understood properly only if we regard it as a passing "from singularity to singularity."[21] This means that the evental emergence of the potentiality "not to" in an example or paradigm retroactively changes what the community of examples given thus far will have been. Any example adds up to the sequence of examples given thus far, as there can never be one final example due to the fact that any example is precisely the singularity being excluded from the set which it stands for and which it constitutes. As a consequence, Agamben can also maintain that the paradigmatic relation oscillates between synchronicity and diachronicity: "it lies neither in diachrony nor in synchrony but rather in the crossing of the two,"[22] that is, it lies between a singularly given paradigm – adding a new singular content – and that which the community of examples will have been – after having added the new singular content to the series of examples obtained thus far. The community of examples is a community of singularities that have nothing in common but the fact of being excluded from the set whose universality they constitute by means of this very exclusion. Paradigms thus stand between the evental emergence in language and the retroactive constitution of the unity of the series of paradigms; between the exposure of the "potentiality not to" in a singularity and its retroaction on the whole series of singularities.

•••

The problem here is that, for Agamben, this movement of retroaction does know an end. There are what we could call two kinds of potentialities implied in his theory of the paradigm. If any example retroacts on the series of examples given thus far, there is nevertheless a point at which retroaction has to stop: the eventality of language as such (or the genesis of the transcendental in itself) which philosophy is able to discuss by presenting the logic of the series of examples. If retroaction is an always necessary element of that which Agamben calls potentiality ("not to"), this is due to two things: (1) there is a moment of negation involved in it – the constitutive exception that generates the singularity of the paradigm and provides us with the background for any retroaction; and (2) there is a moment of privation involved in it – which should ensure that there can never be a substantialisation of the series of examples, as it is always possible to add a further paradigm to it. (This means that the series itself presents the principle of pure belonging as there is no attribute or property which would limit what could become a paradigm, a further element of the sequence of examples.) Therefore, the "potentiality not to" is the result of a logical condensation: a constitutive exception that guarantees the non-substantiality of the series of paradigms. But still the retroaction must end at one point because it cannot retroact on the structure of language, that is, pure belonging as such or, as Agamben has it, whatever being as "being-called."[23]

In other words, philosophy is able to present language in its eventality by presenting paradigms (of eventality and thereby paradigmaticity as such) that form a retroactive series. Yet it is precisely this idea that urges Agamben to specify that there is a transcendental point upon which no retroaction is able to retroact. This transcendental is that which enables the potentiality *not to* and is therefore precisely a *potentiality* not to, that is, (the structure of) language as such. This is to say that its emergence, the *arché*, can never be thought; there can never be an evental emergence of language or anything else which could be thought as retroactively preceding language. Therefore, language becomes itself a transcendental without emergence, which, at the same time, stands for, or better *is*, eventality as such, that is, the emergence of paradigms as such. In the end, archaeo*logy* is always a *paradigma*-tology. More precisely, it resolves itself into *paradigmaticity*. This is what remains implicit in Foucault's work and comes to the fore in Agamben's.[24] In this sense, he can conclude that the *arché* is not a chronological event but a "force working in history."[25]

If language is thus a transcendental possibility of evental emergences that is itself without emergence, Agamben then loses precisely what he wanted to think with the notion of potentiality, that is, the combination of the "not to" of negation and the insubstantiality of privation – the non-totalisability of the retroactive effect of a singularity on the series it sustains. It should be evident by now how this outcome necessarily implies a (negative) ontological substantialisation. Unsurprisingly, the ethico-political task par excellence, which is subservient to such a substantialisation, amounts to an *appropriation* of potentiality, "having a privation," or, even more clearly, being "the master of privation."[26] Agamben's politics is thus reduced to a basic ontological motto: "Potentiality, not freedom."[27]

Finally, this also means that language – when it is thought as such, and this is precisely what philosophy is able to do by presenting its most fundamental "structure" as the basis of the potentiality "not to," that is to say, paradigmaticity – becomes that which fully encompasses any singularity. If one should think singularities without any dialectical link to the traditional category of the universal, and if paradigms as singular events of language are precisely the model for how we should conceive of singularities, then we can deduce that language comprises any singularity by having the structure of a coming community in which singularities continue to be added to the thus far given sequence of events of language. But it is important to emphasise another implication here: thinking language as such also amounts to thinking that which comprises whatever singularity, whatever being, as there are no properties or categories of belonging (to language) involved. Or to make it even more explicit: when we think language as such we think (whatever) being. The claim that the eventality of language as such – and not only singular paradigmatic events occurring within it – can be thought by philosophy implies that language, when adequately conceived, directly articulates and exhibits being. More precisely, it articulates the *truth* of being, that is, nothing other than the coming community as that which is able to encompass whatever (singular) being.

On closer inspection, this argument has even more far-reaching conceptual consequences: if language as such expresses that which being qua being is – i.e., the truth of being – then one cannot but equate being with eventality – owing to the fact that thinking language as such amounts to thinking eventality as such. Language is evental and so is being. Being is an arche-event. Therefore, Agamben's ontology is vitalist in a very specific sense, one that further complicates Deleuze's opposition between power without action and act without essence, since, as we have seen, potentiality is in itself essence. In other words, by thinking, starting from singular paradigms, language as such in terms of paradigmaticity and eventality, Agamben in the end intends to think being as evental. Only by thinking language as such as evental is one put in the position of thinking being (as evental). This is precisely what we call Agamben's linguistic vitalism. The latter is, at the same time, hyper-Heideggerian, for two precise reasons: (1) there is a truth of being which can be thought by thinking language as such (as paradigmaticity, eventality) and the only truth that there is is the truth of being; and (2) philosophy can achieve that which, for Heidegger, only poetry was capable of, namely the articulation of this truth. According to Agamben, philosophy can precisely fulfil the task Heidegger assigned to poetry not only by thinking poetically (*dichtend-denken*) but by becoming poetry.[28] Philosophy becomes poetry precisely when it itself takes the form of paradigmaticity, presenting singularities, paradigms, and their relationship.

Let us re-state this crucial point one more time: Agamben's project is both Heideggerian and vitalist. On the one hand, it is primarily, as indicated above, an ontological and not a political enterprise.[29] Agamben is not concerned with rethinking the possibilities or impossibilities of contemporary political action or with proposing a concrete analysis of concrete situations. He is rather reformulating, in a very personal and original manner, the Heideggerian project of conceiving of the truth of being, of its history and forgetting. This is the most consistent explanation for the complete absence of any consideration about political agency or political organisation in his oeuvre. On the other hand, the primacy of ontology over politics – in the guise of conceiving of the becoming poetry of philosophy as the articulation of the truth of being – is supplemented by what we propose to name linguistic vitalism – the insight that language as such is nothing other than eventality, the exhibition (or expression) of being qua being. Interestingly, this stance has one final implication: the only God that can save us, for Agamben, can be interpellated solely by philosophy.[30]

•••

For the most part, Agamben's work amounts to a *critical* speculation on the metaphysical form of life – which he derives from Heidegger and problematically identifies with the history of the Occident. To this extent, as we have seen, his political discourse – the critique of the form of law – is founded upon a vitalist ontology. The reading we offer is reinforced by the fact that

Agamben himself seems at times to recover an unexpected vitalist Heidegger according to whom, for example, "facticity" – and by extension also *Dasein* – "designates 'the character of being [*Seinscharakter*]' and 'the e-motion [*Bewegtheit*]' proper to life."[31]

As soon as we accept this interpretative point of departure it is, however, imperative to stress that Agamben also attempts to think *affirmatively* the possibility of other – non-metaphysical, or better, vitalistically *meta*-metaphysical – forms of life that would enable the promotion of a new positive politics.[32] Thus, it is taken for granted that the dialectical "distinction between meaning and denotation" that linguistically determines the metaphysical form of life is a "historic product," and not an "original and eternal characteristic of human language."[33] It could cease or be actively overcome. At this contradictory (a-logical for metaphysics) level, Agamben is discussing the genesis of the transcendental as that which, by his own admission, cannot be thought. Three important questions delimit such a discussion in an ever clearer way: is the post-metaphysical, or *meta*-metaphysical, speaking being still a member of the species *Homo sapiens*? Are there forms of life that are *not* human? Can philosophy think the form of life as such?

The first issue is addressed especially in *The Open*. Agamben's postmodern presupposition is that, today, the speaking animal can no longer assume his inessential being as a historical task, and that this state of things is epitomised by the failure of Marxism.[34] Currently, what is difficult to establish is whether the humanity of biopolitics which replaced the collective assumption of a historical task with the impolitical mandate of "tak[ing] upon itself [. . .] the total management of its own animality" – exemplified by the genome project, global economy, and humanitarian ideology – "is still human."[35] Having said this, Agamben hints at the possibility of a different – theriomorphic – "economy of relations between animal and human"[36] in post-history, or better in a "fringe of ultra-history,"[37] that would also entail a reconciliation of man with his animal nature.[38]

The second issue openly surfaces in the recent *The Signature of All Things* – beginning with its very title – but is implicitly present throughout Agamben's work. In the second essay of this book, commenting on Paracelsus and Henry More, he returns to the idea of "natural hieroglyphics,"[39] a dimension of language that is irreducible to semiology, semantics, and hermeneutics. At one point, discussing Foucault's *Archaeology of Knowledge*, Agamben goes as far as claiming that there is a "life of signs" (*vita dei segni*) – such as that of the "eye-shaped spot [*macchia in forma di occhio*] on the corolla of the *Euphrasia*" – a basic "function of existence" that belongs to signs as such.[40] The conclusions that are drawn from this statement are just as far-reaching as they are obscure: being, for the very fact of existing, of giving itself, receives, or "suffers," "marks" and "signatures" that always-already orient it towards a certain hermeneutics; ontology as a discourse on being is therefore a discourse on the "passions of being," that is to say, being and its passions must be identified ("existence is a transcendental dissemination in passions, that is, in signatures"; "in truth being is not a concept but a signature").[41] Here, it would clearly seem that not only are there non-human forms of life but also that being is in itself formalised in a sentient manner. What is at stake in these enigmatic passages can be better grasped if one questions the origins of the phrase "form of life." Agamben refers at times to Wittgenstein but we know from Paolo Virno (with whom he edited a journal of the same name in the early 2000s) that another source was influential for their common project, although they did not necessarily endorse it *in toto*: the writings of the vitalist Jungian biologist Adolf Portmann, especially his *Aufbruch der Lebensforschung* (1965). For Portmann, human and non-human forms of life are forms of appearance, or natural self-presentations, that, in expressing the "mysterious" interiority of life as such, cannot be limited to the functions of self-preservation and the preservation of the species:

> The self-formation of an organism (its ontogenesis) does not only produce a system that guarantees the vital functions of self-preservation, but also leads to a demonstration of the particular mode of being of the form of life in question. The meaning of many organic

> apparatuses [such as the eyes on the feathers of a peacock – one of Portmann's favourite examples] is to manifest, to represent to the senses, the form of life to which they belong.[42]

To sum up, being as the being of life would be in itself nothing other than the modal *hic et nunc* of its expressive formation.[43]

Finally, with regard to the third and most crucial question, Agamben seems to be answering it affirmatively whenever, in his recent work, he evokes the Christian notion of "eternal life" (*zōē aiōnios*). In *The Kingdom and the Glory*, he does not hesitate to affirm that the latter is "the name of [the] inoperative center of the human" – of its inessential essence marked by an absence of aim that itself "make[s] the incomparable operativity of the human species possible"[44] – and hence of the "political 'substance' of the Occident" which the different discursive articulations of dialectics – the "bipolar machine" or "bipolar structure"[45] epitomised by theology – attempt to capture within themselves.[46] Here, the in-itself of the form of life (i.e., "eternal life") amounts to the non-relational relation between life and its form (the logos), between inoperativity and operativity – which is what allows Agamben to identify the form of life (*diagōgē*) with the immobility of the Aristotelian prime mover – between impotence and potentiality. From a slightly different perspective, we could say that the relationality of the form of life as such is non-relational since pure immanence can give itself only as its own transcendental expression: as we have seen, being is nothing other than eventality. Consequently, immanence becomes inseparable from transcendence. The best philosophy can do to articulate the extreme reciprocity of life as form and form as life is to avail itself of the Christian discourse on the co-implication of the transcendent God and the immanent Son. Thanks to theology, philosophy is able to think the immanence of transcendence (Christ's incarnation; his redemptive economy) and the transcendence of immanence (God's separation from the world He nevertheless governs) as ultimately deriving from "eternal life" and flowing out into it.[47]

The function of theology in Agamben's philosophical system is at least threefold. Initially, it would seem to stand, together with politics/law and linguistics, as one of the three privileged discourses by means of which philosophy can investigate the metaphysical form of life, its fundamental dialectic of exclusive inclusion. Just as politics and law are understandable from within metaphysics only as the indissoluble articulation of the sovereign and the *homo sacer*, which is transgressed inherently – and thus confirmed – by the (generalisation of the) state of exception,[48] and, in parallel, linguistics reproduces the same self-refuting, yet productive, apparatus in the triad meaning/denotation/performative speech,[49] so theology ultimately amounts to the inextricability of the Kingdom of Heaven and providential *oikonomia* as it is both suspended and revealed in glory.[50]

On this level, the three discourses in question appear to be almost indistinguishable, but on closer inspection we uncover a clear primacy of theology. Linguistics may at times be presented as a sort of meta-social science of the language of metaphysics – of the logos of the metaphysical form of life – that is always presupposed by other (politico/legal, theological) discourses. For instance, theology seems to be sutured to linguistics as a meta-discourse when Agamben analyses St Paul's Messianic notion of faith (*pistis*) and concludes that, in addition to being a self-referential speech whose effectiveness relies solely on its being uttered – professing one's faith in Christ, the incarnated logos – it cannot be limited to performativity but entails, beyond it, a "pure and common power of saying [*potenza di dire*]."[51] In this case, theology should be understood as the internal frontier of linguistics. At times, politics and law are themselves sutured to linguistics, or at least wholly readable through its paradigms, to the extent that one can claim that, in the state of exception, the law suspends its own application only in order to found, in doing so, its validity precisely in the same way as, in the performative, language brackets its denotative value only in order to establish its connection with things.[52] Yet Agamben also carefully delimits the function of linguistics: if, after more than two thousand years of speculation, the science of language has finally been able to grasp the enunciative function of logos, the formal level at which the speaking being seizes language in a concrete act of speech, what nevertheless remains beyond its capacity is an appreciation of "the *ethos* that is produced in this gesture."[53] While Agamben's identification of poetry with a special domain of language in which, on the contrary, the subject experiences such an ethical relation with his speech appears to be almost inevitable given his Heideggerian background, what comes as a surprise is his (ingenious but improbable) tracing of a direct genealogical connection between the origin of modern Western poetry and Christian Messianism: the poem and, in particular, the institution of the rhyme in courtly poetry reproduces the structure of Messianic time, it is the "messianic heritage Paul leaves to modern poetry [...] The history and fate of rhyme coincide in poetry with the history and fate of the messianic announcement."[54] Poetry as one of the two most proper dimensions of language – the other being philosophy – which, as such, remains inexplicable for linguistics, can ultimately be accounted for only in theological terms.

But similarly, and most importantly, politics is itself often sutured to theology in crucial passages of Agamben's thought: Western sovereignty has always already been split between the Kingdom and the Government due to the Trinitarian formulation the Church Fathers introduced to solve the separation between being and praxis they inherited from Aristotle and the late classical world.[55] Although Agamben pre-emptively warns us that "locating government in its theological locus in the Trinitarian *oikonomia* does not mean to explain it by means of a hierarchy of causes, as if a more primordial genetic rank would necessarily pertain to theology,"[56] there is still here an implicit grading system at work, in spite of an ostensible attempt to avoid the establishment of a meta-discourse by means of an intricate network of interdisciplinary cross-references. More precisely, we could speak of a fundamental asymmetry between, on the one hand, theology and, on the other, politics and linguistics. While theology sutures itself to the other discourses exclusively in the here and now of synchronicity – for instance, through Agamben's linguistic (and as such paradigmatic) analysis of the incipit of the *Letter to the Romans* – politics and linguistics are sutured to theology at the level of both synchronicity – Agamben's paradigmatological thought – and genealogical retroactive diachronicity – for instance, by positing that Western governmentality and its fulfilment in the society of spectacle are the inevitable consequences of the failure of secularisation;[57] or that Western poetry is the return of the Messianic repressed by the Church. In other words, linguistics and politics are paradigmatic sciences of singularly given paradigms (such as performative speech acts or the state of exception); yet theology encompasses both of them by standing between the evental emergence of a paradigmatic science and the retroactive constitution of the unity of the series of paradigmatic sciences. The latter can ultimately be obtained exclusively on condition that we enunciate a contradictory (a-logical) hyper-paradigmatic discourse about the unthinkability of the evental emergence of language as such: this is nothing other than the mystery/ministry of the *evangelium vitae*, God's incarnation, the becoming logos of eternal life.

•••

We do not have to lose sight of the fact that, from Agamben's standpoint, most of what we have been discussing so far in terms of his implicit theory of discourses should be located on the level of the *negation* of onto-theological metaphysics, which is far from entailing its overcoming. We could call this level, that of the performative, the state of exception, and glory, "negative inoperativity." Negative inoperativity implies an "isolation" or "separation of inoperativity into a special sphere,"[58] the categorisation of the purely immanent – or, which is the same, purely transcendent – non-relational relation between life and its form, that is, being as such. In other words, negative inoperativity *formalises* the form of life as form of *life*. Let us dwell on glory, which is the kind of negative inoperativity Agamben has analysed in more depth in the last few years.[59] Through glory, the Church divides inoperativity into cult and liturgy, which culminate in doxology, the infinite acclamation with which the blessed will praise God for eternity.[60] In doing so, it turns the de-activation of the providential economy brought about by the final judgement – which is already entirely contained by the first coming of Jesus – into an "economy of economies."[61] It transforms the "revoking of every *bios*," or *form* of life, brought about by the *zōē tou Iesou* as *evangelium vitae*, the appearance of the "threshold" at which the human can finally assume "the impossibility that life might coincide with a predetermined form" – which is to say, the unformalisability of the force of life – into a "permanent condition."[62]

Here, the crucial question for Agamben can only be: "is it possible to think inoperativity outside of the apparatus of glory?" His straight answer in the concluding sections of chapter 8 of *The Kingdom and the Glory* – which are, as he specifies in the Preface, "the hidden center" around which "the book was written"[63] – is affirmative and unhesitatingly points towards the "theme" of eternal life.[64] Glorious life as negative inoperativity should be countered by Messianic life as positive inoperativity; these are the two opposite expressions of *zōē aiōnios*. For Paul, eternal life indicates the special quality of life in Messianic times, and not simply the future condition of the blessed, one which is marked by a "special indicator of inoperativity," the *hos me* ("as not"); the latter directly actualises our inessential essence.[65] For the followers of the Messiah, it is a matter of living the life that we live as if we were *not* living the specific set of events that compose our biography but rather the pure potentiality of the (force of) life for which and in which we live.[66] On this basis, in the section that concludes *The Kingdom and the Glory*, Agamben more openly speaks of subjectivity *tout-court* and the ensuing possibility of founding a new politics – "a sui generis 'praxis'" – as "the live-*ability* of every life," the inoperativity in which "the life that we live is only the life through which we live; only our power of acting and living, our act-*ability* and our live-*ability*." He then concludes that "here the *bios* coincides with the *zōē* without remainder."[67]

For each of the three discourses he privileges (theology, linguistics, politics/law) there corresponds in Agamben an attempt to think a positive inoperativity that would overcome "separation" (i.e., the unveiling of the point of impossibility of the metaphysical form of life). Without going into detail, in addition to Messianism, we can recover this position in the *literary/fictional* figures of Bartleby and K. – from Kafka's *The Castle*[68] – for politics, and in poetry for the science of language.[69] Agamben also insistently speaks of positive inoperativity in terms of "another use," or "new use," the investigation of which should conclude his *homo sacer* long-term project.[70] Rather than isolating inoperativity within a special sphere and thus limiting it to showing the inessential essence of man – pure potentiality without purpose – a new use (for instance, of the body) would imply a different practical articulation of the relation between power (or potentiality) and act. A new use follows from the realisation that the truly inoperative essence of man does not amount simply to inertia, or absence of action, but "allows the very potentiality [*potenza*] that has manifested itself in the act to appear."[71] This is Agamben's further twist on Deleuze's version of vitalism: the form of life is in itself not only a power without action, im-potent exhibition, but the very presentation (or better, "contemplation") of the *power* of impotence (the power of negation) in the act, what an earlier text enigmatically called "the power of not-not passing to the act."[72]

The fact that, at different stages of his work, Agamben repeatedly adopts terms and phrases such as "exhibition," "im-potence," and "passing to the act" to describe the way in which the "new use" might subvert the entrapment of the form of life in negative inoperativity should have alerted those of his readers who are versed in psychoanalysis. The praxis advanced by the new use is unequivocally perverse. Agamben himself comes to this conclusion in one of the few places in which he gives a concrete example of what he means by positive inoperativity: unlike the glorious body of the blessed which is an "ostensive body whose functions are not executed but rather displayed" – the "glorious penis" and "doxological vagina" of the elected are organic and real but "outside the sphere of any possible use" – a new use of the body would entail perversion, the employment "of the organs of the nutritive and reproductive functions" in a way that "turn[s] them – in the very act of exercising them – away from their physiological meaning, toward a new and more human operation."[73] Most importantly, the paradigmatic function of perversion in Agamben's affirmative project cannot be reduced to the search for a different sexuality. The "as not" (*hos me*) of the Messianic and Franciscan man[74] – "Let's live as if we were not Jewish, circumcised, man..." – is itself emblematically perverse; it stands as the epitome of the *je sais bien, mais quand même* that, according to Freudian and Lacanian psychoanalysis, displays the basic foundation of disavowal (*Verleugnung*) as one of the structural ways the subject tries to cope with the void that in-determines the human animal. Perversion is the "as not" with which each one of us *may* answer to privation. From Agamben's perspective, cunnilingus or fellatio and the *usus pauper* of the Franciscans – a "poor" use of worldly goods that subtracts itself from the appropriations guaranteed by civil law[75] – similarly allow the human to contemplate his *power to* act as a power *in* the act, and thus positively determine him as a "master of privation." This is, for the Italian philosopher, the proper space for politics as a new – and as such always to be repeated – anthropogenesis that goes by the name of "an Ingovernable [*un Ingovernabile*]."[76]

•••

Philosophy's ultimate task is to think other possible forms of (human) life as all resulting from the positive inoperativity by means of which *zōē aiōnios*, the form of life as such, can dissociate itself from metaphysics. But Agamben also admits that philosophy can achieve this purpose only in so far as it becomes a *vera religio*, a true religion.[77] On the one hand, he ends *The Kingdom and the Glory* with a symptomatic attack on Heidegger whose late philosophy "appears to pass into religion." Rather than considering the economy of being, its epochal unveiling in a veiling, as a "political mystery" dependent upon the sovereign dialectic of the Kingdom and the Government, the German philosopher would ultimately understand the problem of *Ge-stell* (technology) – philologically related by Agamben to the themes of *oikonomia* and governmentality – as "the ultimate mystery of God."[78] Agamben's overcoming of Heidegger – and his way of tackling metaphysics – would seem to consist here of a replacement of the primacy of theology over ontology and politics with that of politics over onto-theology. It is, therefore, to say the least, remarkable that the conclusion of *The Sacrament of Language*, published only one year later, presents the function of – Agamben's – philosophy as that of a "true religion." This confirms what we have tried to explain earlier in terms of his overall project being a meta-metaphysics – which thus, in a sense, indeed goes beyond Heidegger's final suturing of ontology to theology, but only by radicalising it. Again, in the end, Agamben's paradigmatological thought needs to rely on, or better, directly posit itself as an a-logical hyper-paradigmatic discourse about the unthinkability (or non-formalisable paradigmaticity) of the evental emergence of language as such, the mystery by means of which *zōē aiōnios* gives rise to political ministry, and *sacramentalises* itself in a *form* of life. It is only on this basis that we can make sense of the enigmatic last pages of *State of Exception* where Agamben states that the aim of thinking (positively inoperative) politics is to "show the nonrelation" – that is to say, the mystery of the ministry – of life and law.[79] What lies beneath, and at the same time sustains, any sort of archaeological enterprise is always the "signifying *power* of language," the "event of language" as a "positive force of language" in which the divine being expresses – or contemplates – itself.[80]

Consequently, not only does theology suture the critique of the metaphysical form of life but it also functions as a meta-metacondition – as the precondition of philosophy as a meta-condition – when Agamben addresses meta-metaphysical being as such, that is, being as inextricable from (its) eventality. In short, his theory of the form of life identifies the mission of philosophy with the elaboration of a *theological* linguistic vitalism. The supremacy of ontology over politics conceals a deeper supremacy of theology over ontology, which obviously does not have to abide strictly to the Christian configurations it has assumed during the last two thousand years.[81] Its supreme stuttering spokesman could well be the laicised priest of a new order of poetic perversion, a post-Franciscan version of Bartleby as new a-theological Christ.[82] Agamben himself seems to be hinting in passing at the inherent contradiction of Christian theology, that between Messianism and the Church's confining it into providential economy and eventually glory, in terms of the tension between an un-founded "a-theological" pole (the ministry of the *mystery*, of the logos's lack of foundation; the affirmation of the *eventality* of being as such) and an institutional "theo-alogical" one (the mystery of the *ministry*, of the indistinction between anarchy and government; the isolation of the *being* of eventality as such).[83] Our contention is that his search for a new politics of the "In-governable" makes him thoroughly, and originally, develop the former. Agamben's ontology recuperates politically the a-theological core of Christianity and plays it against its theo-alogical categorisation. Yet, at the same time, he can do this only by transforming philosophy into a theo-alogical discourse, a "*vera* religio" of the transcendentality of eventality.

All this still overshadows what is at stake in a truly universalist politics – and that of the coming community certainly claims to be one. Agamben shares with Badiou's post-Maoist materialist dialectics and Žižek's Hegelian rethinking of dialectical materialism the radical condemnation of the political apathy generated by the generalised state of negative inoperativity (or state of exception) that dominates contemporary society, what Badiou calls "democratic materialism" and the Italian thinker identifies with "the triumph of *oikonomia*" for which "natural life and its *well-being* seem to present themselves as the ultimate historical task of humanity."[84] With Marx and his contemporary legacy, Agamben thinks of politics as a consequence of man's indetermination, his being an animal that is not defined by any specific operation, "a generic being (*Gattungswesen*)."[85] But against Marx he dissociates the "essential inoperativity" that gives rise to politics from any historical task: thus "politics could be nothing else than the exhibition of man's absence of work and almost of his creative indifference to any task." And, most crucially, it is "only in this sense [that politics] can remain integrally assigned to happiness."[86] Again, here Agamben's ontological vitalism makes him favour a politics of in-*determination* as *hyper*-determination (positive inoperativity; the affirmation of the "as not") over the Marxian subsumption of *in*-determination as given only historically – that is to say, retroactively – to *re*-determination. His final emphasis on happiness – which is recurrent and largely unaccounted for in Agamben's work[87] – rather than on universal egalitarian freedom, seems after all to be advancing a new eudaimonistic politics based on an original re-reading of the Aristotelian notions of power and act, which is tailored to the epoch of biopolitics. The obvious problem is that such a politics of happiness remains in its dandyish semi-autism too close to the "well-being" of the contemporary democratic materialism which Agamben himself vehemently denounces.[88] This is the high and, what is more, unnecessary price that the coming insurrection of those who cannot be seen, the a-subjects, has to pay.

notes

1 See, for instance, G. Agamben, *The Kingdom and the Glory* (Palo Alto: Stanford UP, 2011) 62, 114, 230, 284.

2 On this issue, see L. Chiesa, "Giorgio Agamben's Franciscan Ontology" in *The Italian Difference: Between Nihilism and Biopolitics*, eds. L. Chiesa and A. Toscano (Melbourne: Re.press, 2009) especially 149–56.

3 At the same time, Agamben rightly criticises Heidegger for not seeing that metaphysics is itself, in a different sense, already meta-metaphysical, that is, based on a "Voice" as a "pure will to say without saying" (the English translation of this passage is mistaken and renders "puro voler-dire senza dire" as "pure meaning without speech"; *Language and Death* (Minneapolis: U of Minnesota P, 2006) 61). From this perspective, we will then understand Agamben's thought as *meta*-meta-metaphysical.

The Heideggerian program for conceiving of language beyond every *phone* has thus not been maintained. And if metaphysics is not simply that thought that thinks the experience of language on the basis of an (animal) voice, but rather, it always already thinks this experience on the basis of the negative dimension of a Voice, then Heidegger's attempt to think "a voice without sound" beyond the horizon of metaphysics falls back inside this horizon. (Ibid.)

4 *The Kingdom and the Glory* 246, 251; G. Agamben, *Nudities* (Stanford: Stanford UP, 2011) 52.

5 Idem, "The Work of Man" in *Giorgio Agamben: Sovereignty and Life*, eds. M. Callarco and S. DeCaroli (Stanford: Stanford UP, 2007) 2; translation modified; *The Kingdom and the Glory* 246.

6 On how to conceive of the primacy of re-determination over in-determination in politics taking one's departure from Marx, see F. Ruda, "Humanism Reconsidered," *Filozofski Vestnik* (2009); and L. Chiesa, "The Body of Structural Dialectics," *Nessie. Revue numérique de philosophie contemporaine* 6 (2011).

7 The fact that the overcoming of metaphysics would presuppose a sort of substantialist transvaluation of the unthinkable negativity of the speaking (and mortal) being's *in*-determination into the *affirmation* of his non-relational in-*determination* already surfaces in the first programmatic pages of *Language and Death* (1982):

> We will look beyond Heidegger, leaving ourselves open to the possibility that neither death nor language *originally* belongs to that which lays claim to man [ciò che rivendica l'uomo] [...] The faculty of language and the faculty of death: can the connection between these two "faculties", always taken for granted in man and yet never radically questioned, really remain unthought? And what if man were neither *speaking* nor *mortal* [Agamben's emphasis], yet continued to die and to speak? What is the connection between these *essential determinations?* Do they merely express the same thing in two different guises? And what if this connection would really not take place? (*Language and Death* xii; translation modified, our emphasis unless otherwise indicated)

8 See Davide Tarizzo's contribution to the present collection. The essay was originally published as a preface to D. Tarizzo, *La vita, un'invenzione recente* (Rome and Bari: Laterza, 2010).

9 G. Agamben, "La potenza del pensiero" in *La potenza del pensiero. Saggi e conferenze* (Vicenza: Neri Pozza, 2005) 295; our emphasis. A significantly different and reduced version of this essay appeared in English as "On Potentiality" in *Potentialities: Collected Essays in Philosophy*, by G. Agamben (Stanford: Stanford UP, 2000).

10 Idem, "Absolute Immanence" in *Potentialities* 233. Agamben also defines it as a "pure potentiality that preserves without acting" (ibid.).

11 <http://www.entretemps.asso.fr/Badiou/Agamben.htm>.

12 *The Coming Community* (Minneapolis: U of Minnesota P) 9.

13 <http://www.entretemps.asso.fr/Badiou/Agamben.htm>.

14 See *The Signature of All Things: On Method* (New York: Zone, 2009) 31.

15 Ibid. 17.

16 Paolo Virno develops a similar argument in his analysis of the relation between rules and what he calls "diagrams" (see P. Virno, "Jokes and Innovative Action: For a Logic of Change" in *Multitude between Innovation and Negation* (Los Angeles: Semiotext(e), 2008)). See also Pietro Bianchi's reading of this essay in the present collection.

17 *The Coming Community* 10.

18 See G. Agamben, "On Potentiality" in *Potentialities* 177–84.

19 *The Coming Community* 66.

20 Agamben himself refers to the notion of *homo sacer* in terms of a paradigm (see *The Signature of All Things* 31). *Homo sacer* is included into a given socio-political order only by being excluded from it. Or, to use a Žižekian formula, *homo sacer* is included out of a given socio-political order. A paradigm amounts to the inversion of this logic: it is that which excludes itself from a class which it nevertheless constitutes and, more precisely, to which it can belong only by being excluded from it. So a paradigm is that which is *excluded in*.

To take *homo sacer* as a political paradigm means to consider the one who is included out (of the Western-metaphysical apparatus of sovereignty) as somebody who can constitute a new class for which he stands precisely by excluding itself into it. Methodological paradigmaticity is itself nothing other than the excluding in of that which is included out.

21 *The Signature of All Things* 22.

22 Ibid. 31.

23 *The Coming Community* 9.

24 See F. Ruda's critique of Foucault in "Back to the Factory: A Plea for a Renewal of Concrete Analysis of Concrete Situations" in *Beyond Potentialities? Politics between the Possible and the Impossible*, eds. M. Potocnik, F. Ruda, and J. Völker (Berlin: Diaphanes, 2011) 39–54.

25 G. Agamben, *The Sacrament of Language: An Archaeology of the Oath* (Stanford: Stanford UP, 2011) 10.

26 "To have a potentiality, to have a faculty means: having a privation" ("La potenza del pensiero" 284) (see also 289). Agamben also speaks of a kind of appropriation that "appropriates not a thing, but simply impotence [*impotenza*] and impropriety itself" ("The Passion of Facticity: Heidegger and Love" in *Potentialities* 202; translation modified).

27 "La potenza del pensiero" 290. On this point, Agamben can be criticised from a properly Marxian perspective, as it is precisely such a "pure potentiality" that Marx addresses in *Capital* when he speaks about property as a manifestation of the inversion, or, better, perversion, of the relation between having and being, ends and means. For in capitalism, from the standpoint of circulation, at least initially, having a property or wealth seems to be more desirable than its use (see *Capital* (London: Penguin, 1992) 734–38). Or, as Marx already has it in 1844, "the less you *are*, the less you give expression to your life, the more you *have*..." (Karl Marx, *Economic and Philosophical Manuscripts*, in *Early Writings* (London: Penguin, 1992) 361). On the contrary, Agamben considers "having" (a privation) as the purest expression of the immanent relation between life and language.

28 In *Language and Death*, Agamben speaks of a becoming poetry of philosophy that is also a becoming philosophy of poetry:

> Perhaps only a word in which the pure prose of philosophy would intervene at a certain point to break apart the verse of the poetic word, and in which the verse of poetry would intervene to bend the prose of philosophy into a ring, would be the true human speech. (78; translation modified)

More specifically, poetic thought and philosophical poetry would "balance" (*pareggiare*) – and thus reciprocally annul – the *form* of life (i.e., language as the "plurality" of serial negation) and the form of *life* (i.e., language as an "always" that is "an idea of unity" but also an event, a "surprise" – when it is perceived as such from the perspective of the *form* of life qua "habit"). In all truth, Agamben proposes much more than an equalisation since the "always" – "something alive" – contains in itself *both* unity and the event. The kind of suspension invoked here does not balance "form" and "life" but rather gives rise to a linguistic *life* that claims to give itself immediately, beyond dialectics (the "extinguishing of thought" into *a* thought), as language without negation ("true human speech"), as a "having" language, not being spoken by it (ibid. 79–81; translation modified). As should be clear from our general argument, negativity is here only disavowed, not overcome.

29

> *Theōria* and the contemplative life, which the philosophical tradition has identified as its highest goal for centuries, will have to be dislocated onto a new plane of immanence. It is not certain that, in the process, political philosophy and epistemology will be able to maintain their present physiognomy and *difference with respect to ontology.* (G. Agamben, "Absolute Immanence" 239; our emphasis)

30 For a detailed investigation of the notion of salvation in Agamben's thought, see Jelica Šumič's contribution to the present collection.

31 "The Passion of Facticity" 190.

32 This goes together with a problematisation of Heidegger's notion of facticity, which would still underlie an idea of appropriation as the appropriation of the improper. Facticity should instead

be understood in terms of Messianic "use" (see especially G. Agamben, *The Time that Remains: A Commentary on the Letter to the Roman* (Stanford: Stanford UP, 2005) 34).

33 See *The Sacrament of Language* 55.

34 See G. Agamben, *The Open: Man and Animal* (Stanford: Stanford UP, 2004) 75–77. See also idem, "Heidegger e il nazismo" in *La potenza del pensiero* 336 and "The Work of Man" 6.

35 *The Open* 77.

36 Ibid. 3.

37 Ibid. 12.

38 See Ibid. 3.

39 *The Signature of All Things* 36.

40 Ibid. 63–64.

41 Ibid. 66. The highly tentative status of these arguments is confirmed by the fact that, ten pages later, Agamben blatantly contradicts himself: "Transcendentals are not concepts but signatures and 'passions' of the *concept* of 'being'" (76; our emphasis).

42 A. Portmann, *Le forme viventi. Nuove prospettive della biologia* (Milan: Adelphi, 1989) 268.

43 See, for instance, ibid. 108. Portmann's critique of experimental morphology anticipates *mutatis mutandis* Agamben's linguistic unveiling of the limits of semiology, semantics, and hermeneutics:

> Experimental morphology generally considers its objects as given and confines itself to an investigation of the relations between the system of hereditary factors in the germ and the structures that result from it [...] The form is the system of reference for analysis, but the interpretation of the form itself lies outside experimental morphology. (Ibid. 79)

44 *The Kingdom and the Glory* 246.

45 Ibid. 229.

46 "What is at stake is the capture and inscription in a separate sphere of the inoperativity that is central to human life" (ibid. 245).

47 For Agamben, the philosopher who has developed this argument to its utmost consequences without any (explicit) reference to Christian theology is Deleuze:

> Insofar as immanence is "the movement of the infinite" beyond which there is nothing, immanence has neither a fixed point nor a horizon that can orient thought; the "movement has engulfed everything" and the only possible point of orientation is the vertigo in which outside and inside, immanence and transcendence, are absolutely indistinguishable. ("Absolute Immanence" 226)

48 Today, on the one hand, we are perhaps all virtually *homines sacri*:

> Sacredness is a line of flight still present in contemporary politics, a line that is as such moving into zones increasingly vast and dark, to the point of ultimately coinciding with the biological life itself of citizens. If today there is no longer any one clear figure of the sacred man, it is perhaps because we are all virtually *homines sacri*. (G. Agamben, *Homo Sacer: Sovereign Power and Bare Life* (Stanford: Stanford UP, 1998) 114–15)

On the other hand, we witness a reciprocal strengthening of the state of exception and sovereignty:

> President Bush's decision to refer to himself constantly as the "Commander in Chief of the Army" after September 11, 2001, must be considered in the context of [the] presidential claim to sovereign powers in emergency situations. If, as we have seen, the assumption of this title entails a direct reference to the state of exception, then Bush is attempting to produce a situation in which the emergency becomes the rule, and the very distinction between peace and war (and between foreign and civil war) becomes impossible. (G. Agamben, *State of Exception* (Chicago: U of Chicago P, 2005) 22)

49 In speech acts such as "I swear," "the connection between words and things is not of a semantic-denotative type but performative." The "performative is constituted always by means of a *suspension* of the normal denotative character

of language" and yet, at the same time, "in the performative, language suspends its denotation precisely and solely to *found* its existential [*esistentivo*] connection with things" (*The Sacrament of Language* 55–56; our emphasis).

50

> The process of reciprocal glorification between father and son [. . .] is so intimate that the glorification cannot be said to be produced by the son but only *in* the son. At this point it is clear why the "economy of passion" [. . .] is able to coincide perfectly with the glorious economy through which the son reveals the father. (*The Kingdom and the Glory* 207)

> for this reason, doxology, despite its apparent ceremonial fixity, is the most *dialectical* part of theology. (Ibid. 208; our emphasis)

However, at the same time

> theology never manages truly to get to the bottom of the fracture between immanent trinity [father] and economic trinity [son] [. . .]. This is demonstrated in the very glory that was supposed to celebrate their reconciliation. It is marked by a fundamental dissymmetry in which only the economic trinity is completed at the end of days, but not the immanent trinity. (Ibid. 210)

51 *The Time that Remains* 137. For a more detailed exposition of this argument, see L. Chiesa, "Giorgio Agamben's Franciscan Ontology" 160–61. In *The Kingdom and the Glory*, Agamben defines doxology – i.e., that which "captures" the Messianic element of Christianity – as a "pure will to discourse" (236). The difficulty of distinguishing Agamben's own meta-metaphysical project from his critique of metaphysics is evident here – as it is if we compare the Messianic "pure and common power of saying" with the "pure will to say without saying" which, as we have seen, he attacks in *Language and Death*. Agamben replaces the fundamental voluntarism of the modern subject of metaphysics with a supposed pure potentiality of/for language as such.

52 See *The Sacrament of Language* 56.

53 Ibid. 71.

54 *The Time that Remains* 87. See also 78.

55 Ultimately, for Agamben, Aristotle is the theo-political beginning of the West. This is a constant throughout his work. It is possible to say that the (meta-metaphysical) aim of Agamben's own project is a bio-theo-political re-writing of Aristotle's philosophy beyond metaphysics.

56 *The Kingdom and the Government* xi.

57 Ibid. 4, 255–56, 286–87. On the limits of Agamben's analysis of secularisation and its alleged failure, see Alberto Toscano's contribution to this collection.

58 *Nudities* 100–01. See also *The Kingdom and the Glory* 245, 214 ("The Church formalizes glorification").

59 The almost perfect overlapping of glory and the state of exception, their co-belonging to the same bi-polar apparatus as externally included into it, is one of the unspoken conclusions we should draw from the last part of *The Kingdom and the Glory*. According to Agamben, today's society of spectacle and of the government by consent is the result of a generalisation of glory – of its subsumption under the economy; in modernity, the celestial "economy of economies" is progressively short-circuited with *economics* – that goes hand in hand with today's generalisation of the state of exception (which he investigated in previous books). The *doxa* of opinion and statistics is nothing other than the fulfilment of doxology.

60 See *Nudities* 101–02.

61 *The Kingdom and the Glory* 205.

62 Ibid. 248–49. "Life, which rendered all forms inoperative, itself becomes a form in glory" (*The Kingdom and the Glory* 249).

63 Ibid. xiii.

64 Ibid. 247.

65 Ibid. 248.

66 Ibid. 248–49. The Messianism of the "as not" is at the same time pure (vital) "affirmation" as opposed to "the empty rotation" of glory (ibid. 232).

67 Ibid. 251.

68 However, we should keep in mind that, for Agamben, "Kafka is the greatest theologian of the twentieth century," and his characters should therefore be interpreted accordingly (see Agamben's contribution to this collection).

69 See also Jelica Šumič's and Boštjan Nedoh's contributions to this collection. One could speak of a slight shift in Agamben's understanding of the relation between poetry and philosophy over the last three decades: if in *Language and Death*, as we have seen, he focuses more on the becoming poetry of philosophy and the becoming philosophy of poetry as the paradigmatically human experience of "true speech," in more recent works, such as *The Sacrament of Language*, philosophy is distinguished from poetry, and rather works as a metacondition that thinks poetic inoperativity (see *The Sacrament of Language* 59).

70 See *The Kingdom and the Glory* xiii.

71 *Nudities* 102. "Rather, at stake here is the rendering inoperative of an activity directed toward an end, in order to dispose it toward a new use, one that does not abolish the old use but persists in it and exhibits it" (ibid.).

72 *La potenza del pensiero* 294. See also 289 and *The Kingdom and the Glory* 251.

73 *Nudities* 98–99, 102; translation modified.

74 To the best of our knowledge, the Franciscan (Messianic) *usus pauper* is the only other concrete example of the "new use" offered by Agamben (see *The Time that Remains* 27).

75 For instance, this can be achieved by immediately distributing to the indigent the gifts received from the rich (as beautifully depicted in a scene from Rossellini's film *Francis. God's Jester*).

76 *The Kingdom and the Glory* 65. See also 251. Agamben had already articulated the possibility of a perverse (fetishist) overcoming of metaphysics in "The Perverse Image," the last part of *Stanzas* (1977). In what is one of the very few passages in his oeuvre in which he directly addresses Lacan, he comes to the conclusion that the "gesture of the fetishist [...] succeeds in appropriating an unconscious content without bringing it to consciousness." For this reason, "the analyst can perhaps learn something from the pervert" (G. Agamben, *Stanzas: Word and Phantasm in Western Culture* (Minneapolis: U of Minnesota P, 1993) 146–47). Agamben seems to suggest that such a gesture would effectively counter metaphysics in so far as it would undo the dichotomy between the signifier and the signified. The similarity between this logic of unconscious appropriation and that of the "mastery of privation" we discussed above should be evident. From a Lacanian perspective, however, it is impossible to identify perversion with that which paves the way to the abyss of the signifier (its void), that is, the irreducible non-correspondence between the signifier and the signified. On the contrary, the pervert disavows this abyss in a way that is possibly more "metaphysical" than that of the neurotic's repression.

77 *The Sacrament of Language* 66: "Philosophy [...] must necessarily put itself forward as *vera religio*."

78 *The Kingdom and the Glory* 252–53.

79 *State of Exception* 88.

80 *The Sacrament of Language* 33, 36, 46. See also 50.

81 Having said this, his fascinating discussions of the *Christian* Messianic form of life already offer a solid – albeit questionable – point of reference. Agamben has not yet been able to provide a different – non-Christian – model for his affirmative biopolitics.

82 For Agamben's reading of Bartleby, see "Bartleby, or On Contingency" in *Potentialities*. Deleuze already defines Bartleby as "the new Christ" in "Bartleby; Or, The Formula" in *Essays Critical and Clinical* (Minneapolis: U of Minnesota P, 2007) 90. In order to fully grasp Agamben's (contradictory) model of political subjectivity, shouldn't we think of a Franciscan figure who would act *as if* he or she were *not* a Franciscan? A good example of this may be found in the character of the senile priest in Rossellini's wonderful *Francis. God's Jester*: he is not allowed to fulfil his vocation, i.e., preaching and practising the *usus pauper*; he just cooks for the other priests. Linguistically, he carries out an extreme de-activation of the very language the Franciscans (and *in primis* Francis himself) speak: he does not answer (like Bartleby) but simply repeats confusedly *some* of the words uttered by his brothers, thus rendering their sentences redundant and barely intelligible. He is the real jester of the film. The very idea of becoming a "jester" of the divine perfectly captures Agamben's idea of being as expressive eventality.

83 *The Kingdom and the Glory* 239.

84 "Heidegger e il nazismo" 338; our emphasis.

85 "The thought of Marx [...] seeks the realization of man as a generic being (*Gattungswesen*)" ("The Work of Man" 6).

86 "Heidegger e il nazismo" 338–39.

87 See, for instance, "The Work of Man" 2.

88 In the wake of Pasolini, Agamben's critique of democratic materialism also implies an uncovering of its structural link with a (new) form of – hedonistic – Nazism (see especially "Heidegger e il nazismo" 337–38). With regard to Agamben's treatment of Beau Brummell and dandyism, in this context, he openly draws a parallel between liberation and perversion (see *Stanzas* 53–54).

Index

Note:
Page numbers followed by 'n' refer to notes